Urban Sustainability

Jalil Shaeri
Department of Architecture
Tarbiat Modares University
Tehran, Iran

Ali Cheshmehzangi
School of Architecture, Design
and Planning
The University of Queensland
Brisbane, QLD, Australia

ISSN 2731-6483 ISSN 2731-6491 (electronic)
Urban Sustainability
ISBN 978-981-95-8504-5 ISBN 978-981-95-8505-2 (eBook)
https://doi.org/10.1007/978-981-95-8505-2

This Springer imprint is published by the registered company Springer Nature Singapore Pte Ltd.
The registered company address is: 152 Beach Road, #21-01/04 Gateway East, Singapore 189721, Singapore

We collectively dedicate this book to the three Iranian islands of Abu Musa, Greater Tunb (Tunb-e Bozorg), and Lesser Tunb (Tunb-e Koochak), situated at the very heart of the Persian Gulf.

This dedication affirms their historical, geographical, and political belonging to Iran—unequivocally and without hesitation.

Preface

Negotiations between form and function, aesthetics and climate, and density and liveability have long been a part of the process of designing the built environment. However, the invisible but potent movements of heat, air, and pollution must now more than ever be included in that discussion. Understanding the flow of our surroundings, how wind twists through hallways, how heat lingers between façades, and how fresh air moves through congested urban canyons, is becoming essential as cities grow and the climate changes. It is essential. In this situation, computational fluid dynamics (CFD) becomes an essential design ally rather than a distant and technical luxury.

CFD has been mostly used in the fields of mechanical engineering, aerospace, and advanced industrial applications for many years. However, it is time to change its center of gravity. Today, planners, architects, and urban designers are on the edge of a new paradigm in design, one in which digital environmental simulation is not an afterthought but a necessary tool. That change gave rise to this book. It is a useful manual for people who design districts and infrastructure, envision and sketch buildings, create spaces and systems, and try to develop more ethically and wisely in the face of extreme environmental uncertainty.

CFD provides more than just shapes and statistics, more than numbers and contours that could indeed be mesmerizing. It provides insight. It enables us to model futures before they are constructed, such as how a wind might cool a courtyard, how heat islands might expand or contract, or how air would circulate in a public square or classroom. It identifies vulnerabilities that conventional design approaches frequently miss and helps us find answers to concerns we would not have considered. Furthermore, these insights are becoming essential rather than optional as the physical environment grows more intricate and performance-driven.

However, despite its effectiveness, CFD is still underutilized, misinterpreted, or misused in the urban and construction industries. For many experts, it is too complicated, too ambiguous, and too unrelated to the creative process. This book questions that belief. It makes the case that, given the correct attitude and techniques, CFD may be intuitive, interpretable, and profoundly integrative. It challenges both engineers and designers to transcend the enigma of simulation and pursue a common language where evidence-based design, performance, and aesthetics converge.

In the current context of energy scarcity, indoor air quality issues, climate emergencies, and growing demands for environmental justice, CFD transforms from a technical fix to a strategic instrument. It enables us to inquire: How can we lower energy loads and promote natural ventilation? How can promenades and plazas be made wind-friendly? How can heat stress be avoided in neighborhoods that are susceptible, or how can shade be maximized without sacrificing daylight? These are urgent, real-world design concerns rather than theoretical ones.

This book is also about accessibility. CFD should ot be limited to experts in code silos and lab coats. Practitioners from a variety of fields, including urbanists, architectural engineers, sustainability consultants, and advanced design students, ought to have it in their daily toolbox. It should not terrify but empower. At every level of design, it should also foster a culture of curiosity, iteration, and environmental intelligence.

What lies ahead is promising. CFD is increasingly being included into digital design processes. Real-time feedback loops are being made possible by immersive visualization tools, cloud computing, AI-driven optimization, and parametric modelling. In the future, environmental simulation will be a generating, creative force from the start rather than a post-design validation process. Decisions in that future are influenced by both imagination and physics.

This is not a book that claims to know everything. However, it is a companion for people who are posing more intelligent queries. It is for people who think that good design is about more than simply how something looks; it is also about how it breathes, flows, and adjusts to people and space. It is written with hope, based on real-world experience, and motivated by the conviction that, with the right application, computational tools can assist us in creating ecosystems that are smarter, healthier, and more resilient.

Let this book be a stepping stone in that direction. Let it guide you through the complex yet rewarding terrain of applied CFD and help you see the built world not as static mass, but as a dynamic, living system shaped by forces we can now understand, simulate, and design for. This is just a comprehensive guide for *Applied CFD*, which we believe will be very useful for architects, urban specialists, designers, planners, and anyone who is passionate about making a difference to our built environments.

Tehran, Iran Dr. Jalil Shaeri
Brisbane, Australia Prof. Ali Cheshmehzangi

Acknowledgements We would like to express our sincere appreciation to everyone who helped to develop the ideas of this book on the exciting topic of *Applied CFD*. The authors' combined commitment, research interest, study, and enthusiasm for expanding our knowledge of computational fluid dynamics in architecture and urbanism are reflected in this work. Our perspectives have been enhanced by the knowledge and insights of numerous amazing local and international industries, architects, designers, and urbanists, for which we are incredibly grateful. Above all, we are grateful to our families for their unwavering support and tolerance during this process of learning and cooperation.

Competing Interests The authors have no competing interests to declare that are relevant to the content of this manuscript.

Contents

About the Authors

Jalil Shaeri is a researcher and academic with a specialization in the application of computational fluid dynamics (CFD) to contemporary architecture and urban environmental design. He holds a Ph.D. in Architecture from Tarbiat Modares University, Tehran, where his doctoral work focused on environmental performance modeling and airflow simulation within complex urban and architectural contexts. Over the past decade, Dr. Shaeri has built a multidisciplinary research profile at the intersection of architectural design, urban climatology, and simulation-based performance analysis. His scholarly contributions emphasize the integration of CFD tools into early-stage design processes, the optimization of urban microclimates, and the evaluation of energy-efficient building systems. He has published extensively in peer-reviewed journals and is recognized for translating computational methodologies into practical design strategies. His research is characterized by a strong commitment to applicability, methodological rigor, and cross-disciplinary knowledge transfer. This volume reflects his sustained effort to advance simulation-informed design practices within the fields of architecture and urban planning.

Ali Cheshmehzangi is a Professor and Head of the School of Architecture, Design and Planning (ADP) at The University of Queensland. He is also an Adjunct Professor at Hunan University. He is the Brisbane Olympics Committee member for Games Legacy and Growth Sub-Committee as well as the Expert member of the 2032 Games led by UQ. He has been in the World's top 2 percent field leader since 2021, recognized by Stanford University. He is among the top 20 global scholars in the urban sustainability research area. With a career spanning over two decades, he has made significant contributions to the academic and professional communities, with a focus on sustainable and environmentally conscious design. So far, Ali has published over 500 journal papers, articles, conference papers, book chapters, and reports. He has more than 58 academic books, some of which have received awards at the national, provincial, and municipal levels. He also has received international awards and recognition for his research on urban resilience studies and sustainability research, as well as a 2018 Vice-Chancellor's award for his impactful contribution to higher education.

Chapter 1
From Physics to Practice: Core Fundamentals of Computational Fluid Dynamics (CFD) for the Built Environment

Abstract This chapter introduces the fundamental concepts of computational fluid dynamics (CFD) and highlights its critical role as an analytical tool in architecture and urban environmental design. Positioned as a complement to traditional analytical and experimental approaches, CFD enables the simulation of complex airflow patterns, thermal dynamics, and pollutant dispersion within the built environment. The chapter presents key fluid dynamics principles, including the conservation laws, control volume and differential formulations, material derivatives, and the foundational theories of turbulence and boundary layers. It further explores the spectral characteristics of turbulent kinetic energy and the structure of the atmospheric boundary layer. These theoretical foundations are contextualized through their application at multiple spatial scales—ranging from room-level airflow to building clusters and urban canyons. The chapter concludes with an introduction to computational mesh generation, underscoring the impact of mesh topology on simulation fidelity. Collectively, these elements establish a foundation for the mathematical formulations and numerical modeling techniques discussed in Chapter 2. This progression ensures a coherent understanding of how physical principles translate into computable frameworks for simulating the environmental behavior of urban and architectural systems.

Keywords Computational Fluid Dynamics (CFD) · Conservation laws · Turbulence · Boundary layer theory · Mesh generation

1.1 Introduction

In recent decades, computational fluid dynamics (CFD) has emerged as a transformative tool for the analysis, prediction, and optimization of fluid flow behavior across a wide range of scientific and engineering disciplines. In the fields of architecture and urban planning—where air movement, thermal comfort, pollutant dispersion, and energy performance are critical design concerns—CFD provides a means of integrating rigorous fluid dynamics into the creative and decision-making processes.

J. Shaeri and A. Cheshmehzangi, *Applied CFD*, Urban Sustainability,
https://doi.org/10.1007/978-981-95-8505-2_1

Its ability to simulate real-world scenarios under controlled digital environments allows designers to evaluate performance before construction, reducing both risk and cost. Moreover, CFD facilitates data-informed design strategies that respond to environmental challenges and promote healthier, more resilient built environments [1, 2].

Traditionally, the study of fluid mechanics has relied on two fundamental approaches: analytical modeling and experimental testing. However, the growing complexity of real-world flow scenarios, particularly in the built environment, has exposed the limitations of these conventional methods. The development of advanced numerical algorithms and the expansion of high-performance computing resources have led to the establishment of CFD as a third, equally vital pillar of fluid dynamics [3, 4]. CFD enables engineers, architects, and planners to simulate complex physical phenomena that are otherwise difficult, expensive, or impossible to reproduce through physical testing alone.

This chapter offers a structured introduction to CFD, with emphasis on its relevance to the analysis of the built environment. It outlines the fundamental conservation laws—mass, momentum, and energy—and presents both control volume and differential formulations that serve as the mathematical backbone of CFD solvers. These formulations are reinforced with physical interpretations that bridge theoretical principles with real-world design challenges.

Special attention is devoted to the treatment of turbulence, boundary layer theory, and the structure of the atmospheric boundary layer—key concepts for modeling wind flow around buildings, over rooftops, and within urban canyons. By addressing turbulence scales, energy spectra, and eddy formations, the chapter equips the reader with foundational insight for advanced simulations in microclimatic and environmental modeling. These principles are critical for accurately capturing the dynamic interactions between airflow and built structures, especially in densely populated urban areas. Understanding the vertical stratification and diurnal variability of the boundary layer further enables precise modeling of heat transfer and pollutant dispersion. Emphasis is also placed on the implications of surface roughness, thermal gradients, and urban morphology in shaping localized wind patterns.

The chapter concludes with an introduction to computational mesh generation, comparing structured and unstructured grids, and emphasizing the influence of mesh topology and resolution on numerical accuracy and stability. Throughout the chapter, practical examples are provided—ranging from single-room airflow simulations to urban block-scale studies—to demonstrate the applied value of CFD in architecture and city-scale planning. In doing so, this chapter lays a comprehensive scientific foundation for the advanced numerical methods, turbulence models, boundary condition strategies, and real-world CFD case studies that will be explored in subsequent chapters.

1.2 What Is Computational Fluid Dynamics (CFD)?

CFD is a field of study that employs numerical techniques to obtain approximate solutions to fluid flow and heat transfer problems. The physical behavior of fluids is governed by three fundamental conservation laws:

1. Conservation of mass,
2. Conservation of momentum (Newton's second law), and
3. Conservation of energy.

Conservation of mass, also known as the continuity equation, is the law that ensures mass is neither created nor destroyed within a control volume. In fluid dynamics, it implies that the rate at which mass enters a system is equal to the rate at which it leaves, plus any accumulation within the system. This principle is essential in modeling incompressible and compressible flows and forms the basis for understanding how fluid density and velocity fields evolve over time and space.

Conservation of momentum (Newton's second law) is the law states that the rate of change of momentum of a fluid element is equal to the sum of the forces acting on it. These forces typically include pressure gradients, viscous stresses, and body forces such as gravity [5]. The resulting equations—commonly known as the Navier–Stokes equations—are central to CFD and describe how fluid velocity changes in response to internal and external influences. Accurate modeling of momentum conservation is key to capturing flow acceleration, turbulence, separation, and interaction with solid boundaries.

Lastly, conservation of energy is the principle that asserts the total energy within a fluid system is conserved, accounting for internal energy, kinetic energy, and thermal energy. In CFD, the energy equation governs the distribution of temperature and heat transfer through conduction, convection, and radiation [6, 7]. It is critical for simulations involving thermal comfort, passive design strategies, HVAC performance, and energy exchange in urban microclimates.

Together, these laws form the backbone of fluid flow modeling and are discretized and solved numerically in CFD to predict the dynamic behavior of fluids within architectural and urban contexts. These laws can be formulated as systems of partial differential equations, expressed in both integral and differential forms. CFD seeks to approximate these equations by replacing them with algebraic expressions that can be solved computationally [8]. These algebraic equations are defined over a discretized domain in space and time, producing numerical solutions that represent the evolution of fluid variables (such as velocity, pressure, and temperature). It should be emphasized that solving these equations often requires intensive computational resources, well beyond the scope of manual calculation. With advancements in hardware—including faster processors, larger memory capacities, and parallel computing technologies—CFD simulations can now be conducted with high fidelity over complex geometries and realistic boundary conditions. As such, modern CFD relies on the use of high-performance computing platforms and, in many cases, supercomputers to achieve accurate and efficient results.

1.3 Why Use Computational Fluid Dynamics?

Aircraft are among the most widely employed transportation systems in modern society. Large commercial and cargo airplanes routinely take off, travel at high speeds, and land with precision and reliability. The ability to perform these complex operations is heavily reliant on the principles of CFD. From simulating the internal flow in jet engines to predicting external aerodynamic forces acting on the fuselage and wings, CFD plays a central role in the design and performance assessment of aircraft systems [8, 9].

Traditionally, fluid dynamics has been grounded in two core methodologies: theoretical modeling and experimental testing. For instance, to understand airflow around a wing or fuselage, one might first apply governing equations based on fluid mechanics theory and then validate the predictions through physical experiments—such as wind tunnel tests. With the advent of high-speed digital computing and the development of robust numerical algorithms, it became possible to simulate complex physical phenomena computationally. This gave rise to computational fluid dynamics as a third pillar in fluid flow analysis. CFD effectively bridges the gap between theory and experiment, providing detailed insight into flow fields that may be difficult or costly to obtain through analytical or experimental means alone [6, 10]. It is important to note that CFD is not a substitute for theory or experimentation; rather, it complements both and enables a more comprehensive approach. A balanced integration of theory, experiment, and computation is essential for accurate and reliable fluid dynamic investigations. CFD, in this context, serves as a powerful tool to interpret and extend the results derived from the other two approaches (Fig. 1.1).

1.4 Advantages and Limitations of CFD

The main benefits and drawbacks of computational fluid dynamics (CFD) are depicted in Fig. 1.2. Engineers, architects, and researchers can model complicated physical phenomena using CFD, a sophisticated simulation technique that produces quick and reliable results for a variety of operating situations, without the need for expensive or time-consuming physical prototyping. It is essential for performance-driven design in both architectural and urban environments due to its ability to visualize fluid flow behavior in fine spatial and temporal resolutions.

Additionally, parametric research and scenario testing in controlled digital settings are made possible by CFD, which facilitates design optimization and decision-making in a variety of scenarios. CFD models can be modified and reused after being verified against field or experimental data, providing predictive insight and long-term value. CFD is not without its drawbacks, though. Numerical errors and stability problems may arise when solving the governing equations, particularly when working with complicated geometries, shifting boundaries, or extremely turbulent and transient flows. Mesh quality, turbulence modeling, boundary conditions, and

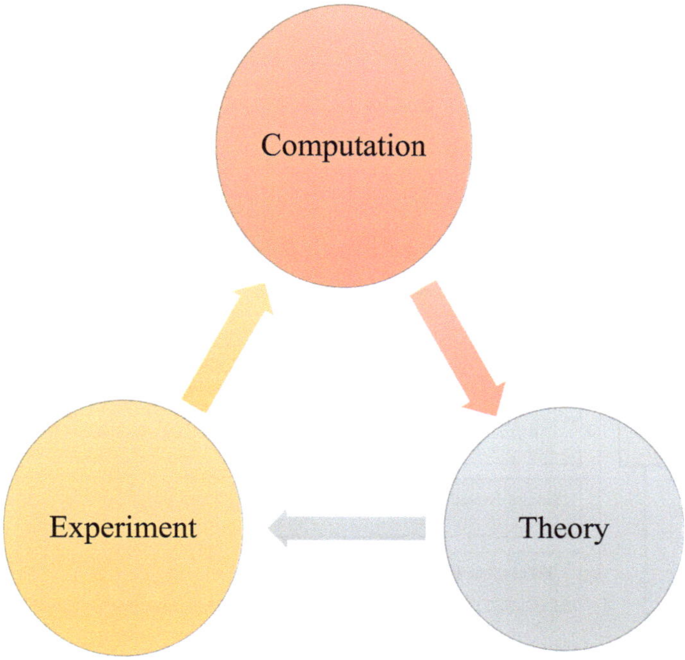

Fig. 1.1 The three foundational pillars of fluid dynamics

solver parameters all have a significant impact on the correctness of the results. The validity of simulations might be jeopardized by misleading outputs caused by poorly defined models or excessively coarse meshes.

Furthermore, high-fidelity simulations can be computationally expensive, especially when dealing with coupled multi-physics analyses or three-dimensional, time-dependent situations. To guarantee convergence, stability, and consistency across simulations and to interpret results in a meaningful way, expertise is needed.

Although CFD provides impressive capabilities for environmental study and design innovation, as shown in Fig. 1.2, users must be mindful of its computational limitations and inherent trade-offs. For CFD to be used responsibly and effectively in practical applications, a balanced awareness of its advantages and disadvantages is necessary.

1.5 Comparison Between Wind Tunnel Testing and CFD

CFD results can be directly compared with experimental data obtained from wind tunnel testing, providing a means for validation and cross-verification. While physical wind tunnels require substantial infrastructure, maintenance, and operational cost, CFD offers a flexible and computationally accessible alternative. Although CFD

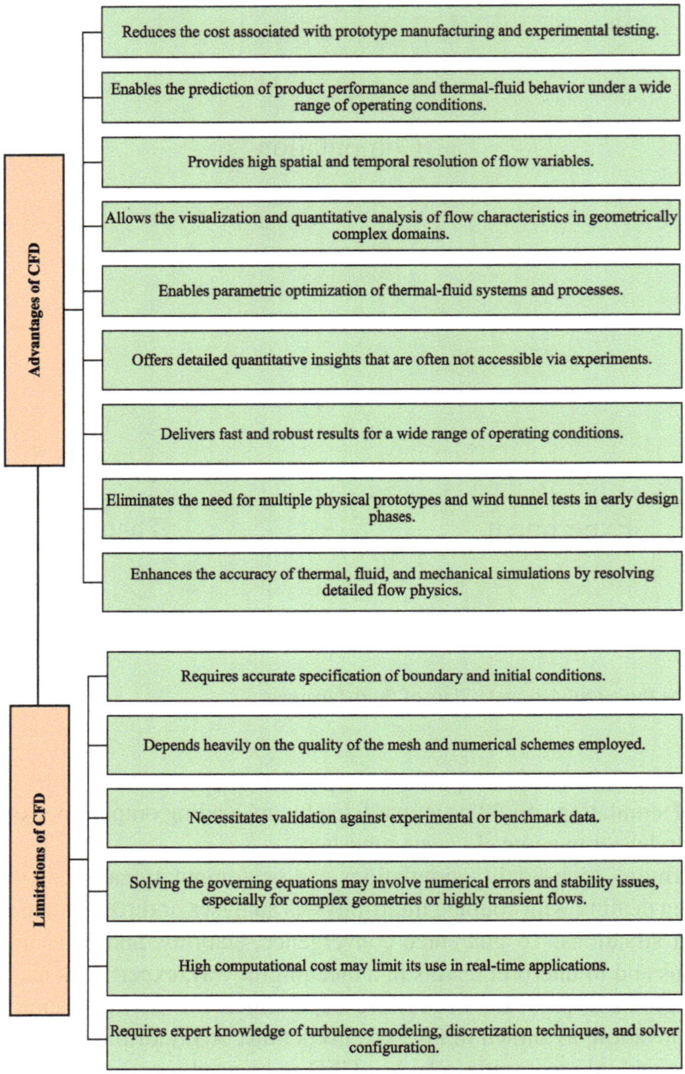

Fig. 1.2 Advantages and limitations of CFD

software may appear effortless to operate on a computer, it still demands careful modeling, expert interpretation, and high computational precision to yield reliable results.

Both methods have distinct strengths: wind tunnels capture real-world physics with fewer assumptions, while CFD provides detailed insight into flow variables that are often difficult or impossible to measure experimentally. In practice, a hybrid approach—using wind tunnel data to calibrate and validate CFD models—can lead to the most robust and comprehensive design evaluations. As computational power and modeling techniques continue to advance, CFD is increasingly complementing, and in some cases replacing, physical testing in early design and optimization phases.

1.6 Applications of CFD

Computational fluid dynamics has found extensive applications across a wide range of engineering disciplines, including aerospace, automotive, civil engineering, and industrial manufacturing. Some representative examples are briefly outlined below.

- **Automotive Industry**

In the automotive sector, CFD is employed to optimize engine combustion, design aerodynamically efficient car bodies, and analyze thermal management systems. Engineers can gain deeper insights into fluid dynamics phenomena and consequently design vehicles with improved performance, fuel efficiency, and safety.

- **Aerospace Engineering**

CFD is widely used in aerospace applications, including the analysis of combustion processes in jet engines, and the aerodynamic performance of aircraft and spacecraft. The ability to simulate complex flow structures around wings, fuselages, and propulsion systems has made CFD an indispensable tool in modern aerospace design.

- **Industrial Manufacturing**

CFD is utilized in various industrial processes, including oil and gas, metallurgy, and petrochemical production. For example, Fig. 1.3 shows a mold-filling simulation where molten metal is poured into a die through upper and lower gates. The velocity field reveals how the molten material propagates over time. CFD enables precise prediction of flow behavior and thermal characteristics, which are crucial for optimizing casting quality.

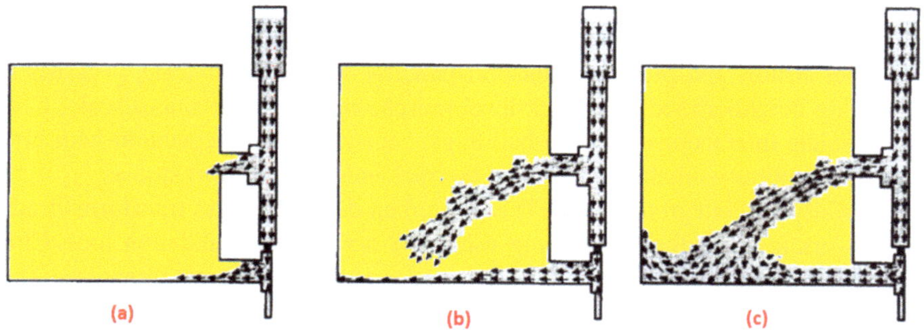

Fig. 1.3 CFD simulation of mold filling in metal casting

- **Application of CFD in Integrated Urban Design**

In recent years, CFD has become an indispensable tool in the domain of integrated urban design, which seeks to unify environmental, social, economic, and cultural dimensions within a coherent spatial planning framework. As urban environments grow more complex and denser, the role of CFD in evaluating environmental performance across multiple spatial scales—from city-wide macro scale to microclimatic local scale—has become increasingly critical.

Integrated urban design emphasizes a holistic approach to shaping the built environment by simultaneously considering factors such as wind flow, solar exposure, pollutant dispersion, and thermal comfort. In this context, CFD functions not merely as a simulation tool, but as an integral component of the design thinking process, guiding informed decision-making during both pre-design and iterative development stages.

The primary applications of CFD in integrated urban design include:

- Urban ventilation analysis, evaluating wind flow between and around buildings to enhance natural ventilation and reduce urban heat island effects;
- Thermal comfort assessment in outdoor public spaces, courtyards, parks, and pedestrian zones under varying solar and wind conditions;
- Pollutant dispersion modeling in street canyons and high-density areas to support air quality management and health-oriented urban planning;
- Optimization of urban form and layout, including the configuration of buildings, open spaces, and vegetation to promote airflow permeability and microclimatic resilience;
- Scenario-based design evaluation, where multiple urban form alternatives are simulated to assess environmental performance prior to finalization of the masterplan;
- Informed positioning of design elements, such as windbreaks, shading devices, tree placement, or open corridors, to enhance spatial and thermal efficiency.

CFD simulations enable urban planners, architects, and environmental designers to visualize and quantify fluid dynamics phenomena that are otherwise difficult to measure or predict in early-stage design. At macro scale, CFD is used to assess city-level airflow and pollutant transport patterns. At the meso scale, it assists in evaluating neighborhood and block-level morphologies. At the micro scale, CFD can simulate interactions between buildings, courtyards, and vegetation with high spatial resolution—suitable for detailed thermal and wind comfort studies.

Critically, the integration of CFD in the urban design process transforms traditional workflows. Rather than being used post-design as a validation tool, CFD contributes dynamically to generating, testing, and refining design alternatives. This approach aligns with contemporary performance-driven urban design practices and enables the development of climate-responsive, resilient, and human-centric urban spaces.

As illustrated in subsequent chapters, the theoretical foundations presented in this chapter—particularly those related to turbulence modeling, boundary layer behavior, and conservation laws—are essential for implementing CFD simulations in realistic urban scenarios. Together, these elements establish the scientific basis for integrated environmental analysis in architectural and urban design projects.

Figure 1.4a illustrates how CFD tools are employed during the detailed design phase of a semi-open courtyard within an urban setting. In this context, design decisions regarding activity patterns, circulation flows, and land-use allocations are informed and shaped by the outcomes of environmental performance analysis [10]. Figure 1.4b illustrates an example of the output of air velocity and pressure of a part of the coastal context in the ANSYS Fluent software.

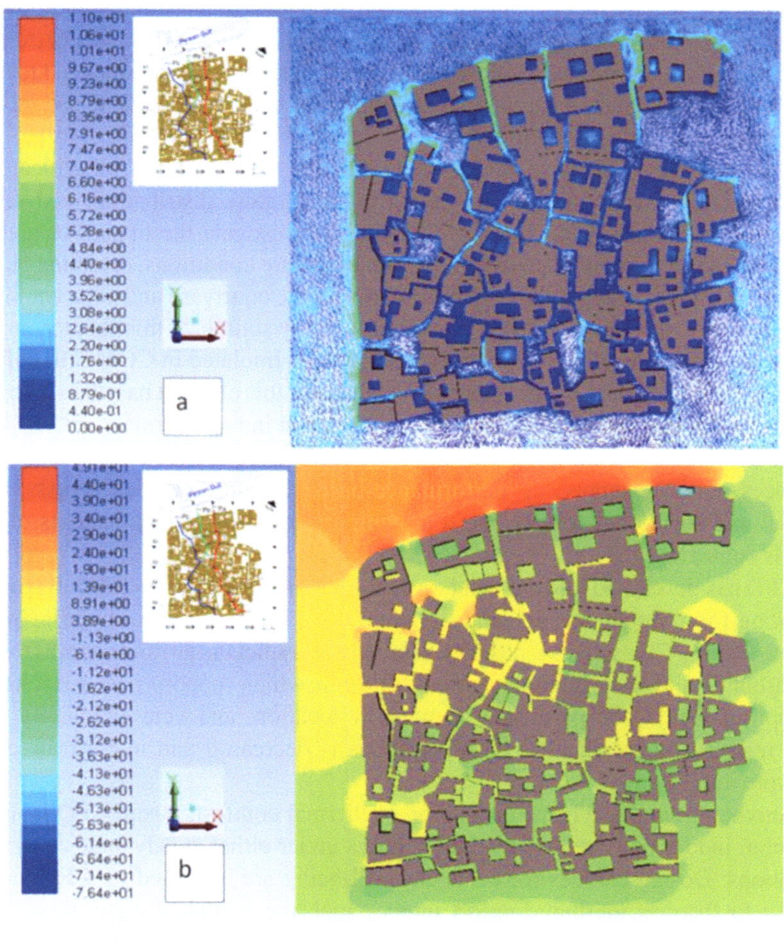

Fig. 1.4 Using CFD in urban design, The pressure (Pa) and air speed (m/s) at 1.2 m above the ground for the selected case study by CFD analysis of ANSYS Fluent

Simulations can quantify how different building forms influence wind speed, or determine which urban configurations are best suited to local climate conditions for enhancing outdoor comfort. Additionally, software such as ENVI-met can be used to analyze urban heat islands and evaluate how different types of vegetation affect temperature and humidity, aiding in the selection of optimal tree species for urban greening.

- **Architecture and Building Design**

CFD is increasingly applied in the fields of architecture and building engineering to simulate airflow and thermal behavior within and around buildings. One of its key applications is in the analysis and optimization of HVAC systems, including the placement of heating and cooling elements such as radiators, air vents, and convectors [10]. CFD is also used to evaluate the performance of passive architectural features, which are essential in vernacular and sustainable building design [9, 11]. For example, windcatchers, underground cooling spaces, and courtyards can be analyzed to assess their effectiveness in enhancing natural ventilation, modifying airflow patterns, and improving indoor thermal comfort.

By simulating airflow behavior, CFD enables the prediction of critical environmental parameters such as air velocity, temperature distribution, and relative humidity. In one study, CFD was employed to investigate the impact of a central water basin in a traditional courtyard on microclimatic conditions, revealing its positive effect on increasing relative humidity in both the courtyard and adjacent interior spaces as shown in Fig. 1.5a [12]. In another study by authors of this chapter, an innovative passive element called a wind chimney was simulated in COMSOL software (see Fig. 1.5b). The aim was to analyze the impact of this element based on innovative design on natural ventilation inside the building and indoor thermal comfort [11].

CFD has become an essential analytical tool in the architectural and building sciences domain, supporting performance-based design strategies across various spatial and functional scales. The following are the primary applications of CFD in this context:

CFD has developed into a crucial tool for environmental and architectural design, allowing for in-depth research and simulation of a broad variety of building performance factors. CFD makes it possible to precisely predict the airflow patterns through and around buildings in natural ventilation design, which supports passive techniques by optimizing façade permeability, window location, and vent sizing. By doing this, reliance on mechanical HVAC systems is decreased and interior air quality is improved.

Another crucial use is the evaluation of thermal comfort, where CFD is used to examine indoor and outdoor thermal settings under either steady-state or dynamic situations. Design choices for occupant well-being are informed by CFD models' ability to forecast thermal comfort indices such the predicted mean vote (PMV) and the predicted percentage of dissatisfied (PPD), which take into consideration convective heat transfer, solar radiation, and occupant-generated heat gains.

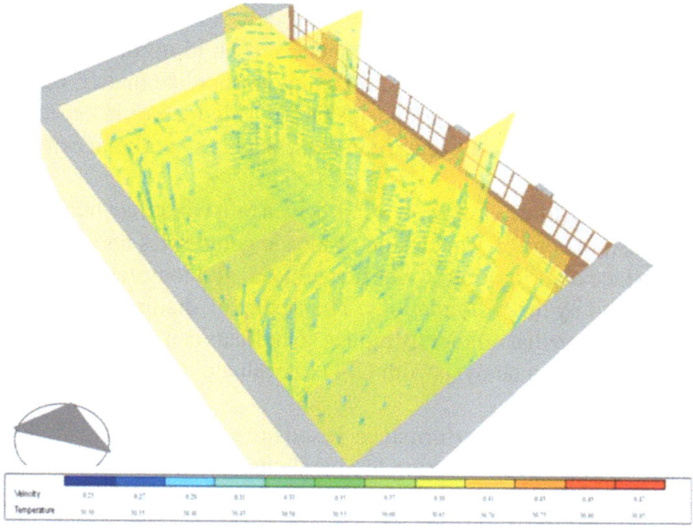

(a)

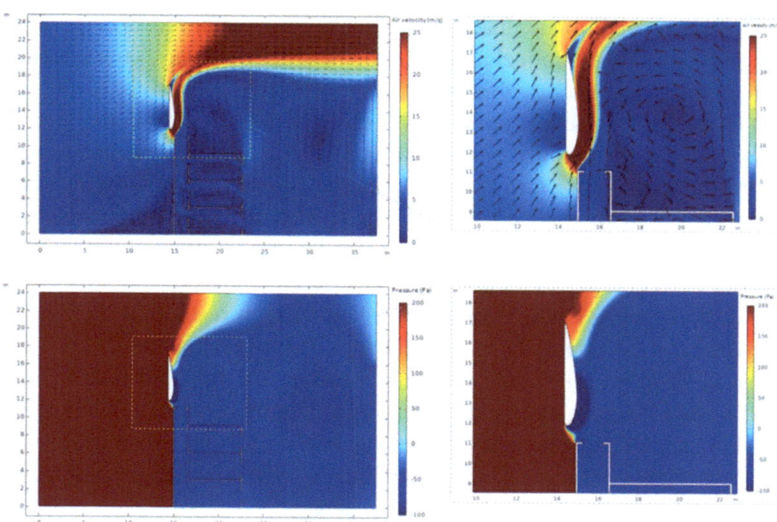

(b)

Fig. 1.5 Using CFD in building design; **a** Indoor airflow velocity and room temperature distribution in the historical Montazeri-Nejad House, Shiraz by CFD of DesignBuilder software; **b** Air velocity (up) and Pressure (down) of building with wind chimney

Additionally, CFD is essential for assessing indoor air quality (IAQ) and pollutant dispersion. It can replicate the movement and build-up of airborne pollutants in enclosed and semi-enclosed spaces, including particulate matter, carbon dioxide (CO_2), and volatile organic compounds (VOCs). This feature is especially helpful in sensitive or high-occupancy environments, such as workplaces, schools, and hospitals.

CFD models aid in forecasting how wind-driven rain will affect building envelopes in terms of moisture control. The design of façade geometries and cladding systems that reduce surface deterioration, water penetration, and long-term moisture-related problems is guided by these simulations. The design of shading devices, overhangs, and façade articulation that efficiently reduce solar loads can also be aided by the integration of solar radiation analysis with CFD to evaluate heat gains through building surfaces.

CFD is frequently used in performance-based fire safety design to model how heat, smoke, and toxic gases behave during fire incidents. The design of evacuation plans, smoke control systems, and fire compartmentation in intricate spatial arrangements are all supported by these simulations. CFD models temperature stratification, airflow distribution, and overall ventilation efficacy to evaluate HVAC system performance in mechanically conditioned spaces. This makes it possible to arrange diffusers, returns, and thermal zoning techniques in the best possible way, guaranteeing both occupant comfort and energy efficiency.

CFD models assess pedestrian-level wind comfort and safety in the urban interface, especially in the vicinity of building perimeters, open plazas, and transitional public areas. Criteria for wind comfort, like the Lawson or NEN 8100 standards, are frequently used to evaluate the usefulness and safety of outdoor spaces. By integrating with building performance systems like EnergyPlus or TRNSYS, CFD further improves energy simulation. This integration enhances the predicted accuracy of whole-building energy models by enabling high-resolution spatial study of convective heat transfer and thermal boundary condition refinement.

Lastly, CFD plays a key role in comprehending the intricate flow behavior in atria and façade systems, particularly in designs that incorporate tall empty spaces, double-skin façades, and buoyancy-driven ventilation. In high-rise and large-volume design, these studies help create climate-responsive envelopes that encourage natural airflow, daylight penetration, and energy efficiency.

1.7 Finite Control Volume

In fluid mechanics and thermodynamics, the control volume concept allows analysis to be confined to a specific portion of the flow domain rather than the entire field. A control volume, denoted by V, is a finite region in space through which fluid flows. The control surface, denoted by S, is the closed surface enclosing the control volume.

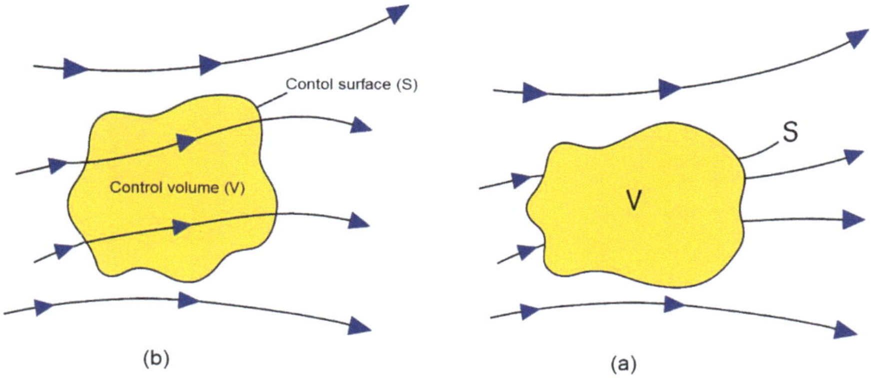

Fig. 1.6 Control volume (V) and control surface (S) in two cases, **a** the control volume moves along with the flow and **b** the control volume is fixed in space while the fluid flows through it

Two distinct configurations are typically considered:

- A stationary control volume through which fluid flows (Fig. 1.6b),
- A moving control volume that travels along with the fluid (Fig. 1.6a).

When fundamental physical principles (e.g., mass, momentum, and energy conservation) are applied to a control volume, they yield integral forms of the governing equations. These integral equations can be subsequently transformed into partial differential equations (PDEs) through limiting processes.

- If the control volume is fixed in space, the resulting equations—whether integral or differential—are referred to as conservative formulations.
- Conversely, when the control volume moves with the fluid, the resulting equations are termed non-conservative formulations.

These formulations are foundational to both analytical and numerical treatments in computational fluid dynamics (CFD).

1.8 Infinitesimal Fluid Element

An alternative to the control volume approach is the infinitesimal fluid element formulation, wherein the fundamental physical laws are applied directly to a differential fluid parcel. This approach leads inherently to the differential form of the governing equations.

Let $\partial\vartheta$ represent the infinitesimal volume of a fluid element within the flow field. Although the element is infinitesimally small, it is sufficiently large to encompass a statistically meaningful number of fluid molecules. The symbol $\partial\vartheta$ corresponds to the differential volume element.

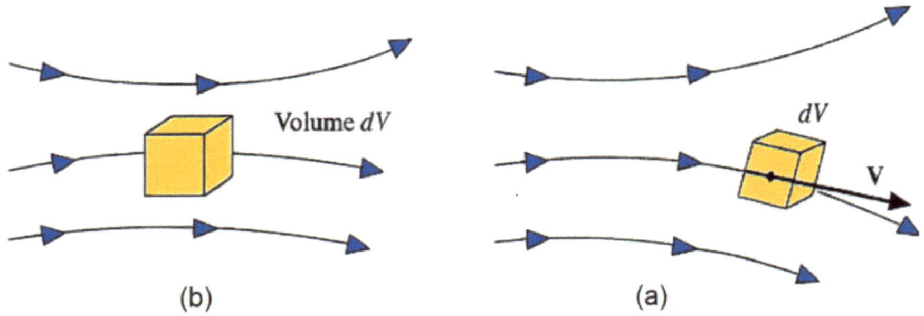

Fig. 1.7 An infinitesimal fluid element in two cases, **a** the control volume moves with the fluid flow and **b** the control volume is fixed in space and the flow passes through it

As with control volumes, two configurations are considered:

- A fixed fluid element, through which flow passes (Fig. 1.7b),
- A moving fluid element that travels with the fluid motion (Fig. 1.7a).
- Governing PDEs derived for fixed fluid elements are referred to as conservative equations.
- PDEs derived for fluid elements moving with the flow are termed non-conservative equations.

This framework underpins the derivation of the continuity, momentum, and energy equations and is a cornerstone of differential-based CFD formulations.

1.9 Material Derivative (Instantaneous Rate of Change with Time)

In Fig. 1.8, a fluid particle, or an infinitesimal fluid element, moving with the flow is adopted as the reference frame. The standard Cartesian unit vectors along the x-, y-, and z-directions are denoted by i, j, and k, respectively. The flow is assumed to be smooth, continuous, and differentiable, and the velocity field $\mathrm{V} = V(xyzt)$ is considered as a function of both space and time.

The material derivative, also referred to as the substantial derivative, represents the total rate of change of a scalar or vector quantity as experienced by a moving fluid particle. It accounts for both the local (temporal) rate of change and the convective (spatial) rate of change due to the motion of the particle through a non-uniform field. The operator is expressed as:

$$\frac{D}{Dt} = \frac{\partial}{\partial t} + \mathrm{V} \cdot \nabla \tag{1.1}$$

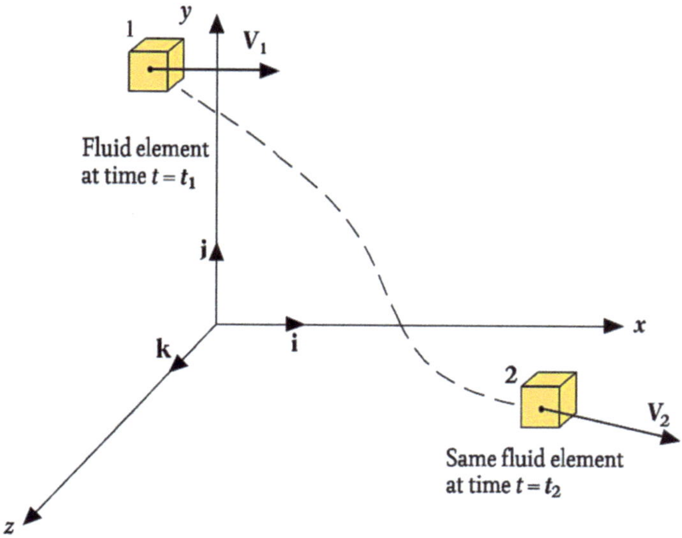

Fig. 1.8 Material derivative of an infinitesimal fluid element in the case where the control volume moves with the fluid flow

This operator plays a central role in the derivation of the governing equations of fluid dynamics, including the conservation of mass, momentum, and energy.

1.10 Velocity Divergence and Its Physical Interpretation

To understand the divergence of velocity, consider a control volume moving with the fluid flow, consisting of the same set of fluid particles. Under the assumption of constant mass, the fluid within the control volume is conserved.

However, as this volume enters regions of spatially varying density, its shape and surface area evolve continuously, leading to a change in volume over time (see Fig. 1.9). The rate of volume change of such a deforming control volume is expressed by:

$$\frac{D\vartheta}{Dt} = \frac{1}{\Delta t}\iint (V\Delta t)\cdot ds = \iint V \cdot ds \tag{1.2}$$

where

- $\frac{D\vartheta}{Dt}$ is the rate of change of the control volume with time,
- ds is an infinitesimal oriented surface element on the control surface,
- V is the local velocity vector, and
- Δt is a finite time interval.

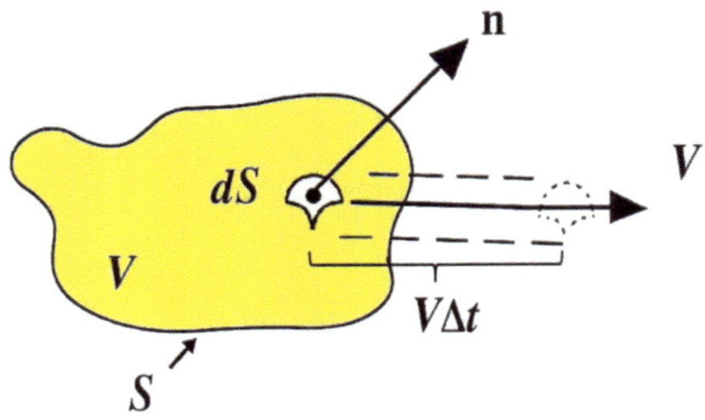

Fig. 1.9 Schematic representation of velocity divergence

The geometric interpretation of this relation is that the volume change $\Delta\vartheta$ of the control volume during Δt is equivalent to the volume of a cylindrical segment whose base is ds and height is given by $(V\Delta t) \cdot n$, where n is the unit normal vector to the surface element ds.

This formulation is essential in interpreting divergence as the net volumetric outflow per unit volume and leads directly to the continuity equation in differential form.

1.11 Three Fundamental Principles Governing Fluid Flow

A rigorous understanding of fluid dynamics is built upon three core conservation laws: mass conservation (continuity), momentum conservation, and energy conservation. These principles are used to derive the governing equations that describe the behavior of fluid systems. Together, they form the foundation for modeling fluid motion in both natural and built environments. In the context of computational fluid dynamics (CFD), these equations are discretized and solved numerically to simulate complex flow phenomena. Each conservation law captures a different aspect of fluid behavior—ensuring mass balance, accounting for force interactions, and tracking thermal energy exchanges. Mastery of these laws is essential for producing accurate, reliable simulations in architectural and urban applications.

1.11.1 Principle of Mass Conservation (Continuity)

The continuity equation expresses the conservation of mass in a fluid flow. This principle can be interpreted in several ways. One common approach involves a fixed control volume through which fluid flows, as illustrated in Fig. 1.10.

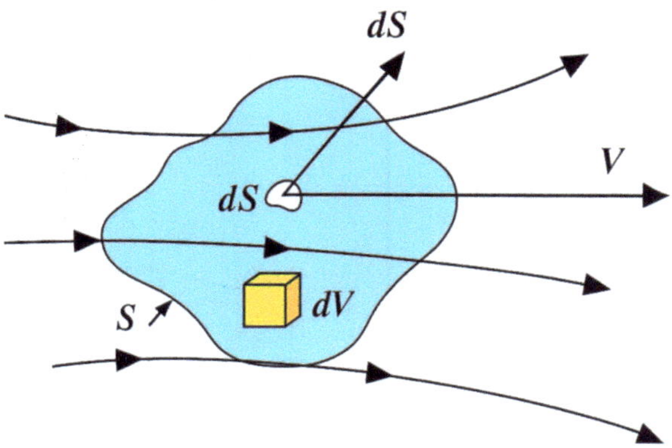

Fig. 1.10 Schematic of a fixed control volume model in space illustrating the principle of continuity

Let V denote the fluid velocity vector, $d\mathrm{S}$ the oriented differential surface vector, and dV an infinitesimal volume element inside the control volume. The integral form of the continuity equation over a fixed control volume ϑ is given as:

$$\frac{D}{Dt}\iiint\limits_{\vartheta} \rho d\vartheta + \iint\limits_{S} \rho \mathrm{V} \cdot d\mathrm{S} = 0 \tag{1.3}$$

This equation states that the rate of mass accumulation inside the control volume is equal to the net mass flux out of the control surface.

Alternatively, by choosing a control volume that moves with the fluid (a material volume), as shown in Fig. 1.11, and applying the conservation of mass to an infinitesimal fluid element, the differential form of the continuity equation is derived:

$$\frac{D\rho}{Dt} + \nabla \cdot (\rho \mathrm{V}) = 0 \tag{1.4}$$

Equation 1.4 indicates that any change in density observed by a moving fluid element is balanced by the divergence of mass flux.

1.11.2 Principle of Momentum Conservation

The momentum equation is derived from Newton's Second Law, which states that the rate of change of momentum in a fluid element is equal to the sum of external forces acting on it.

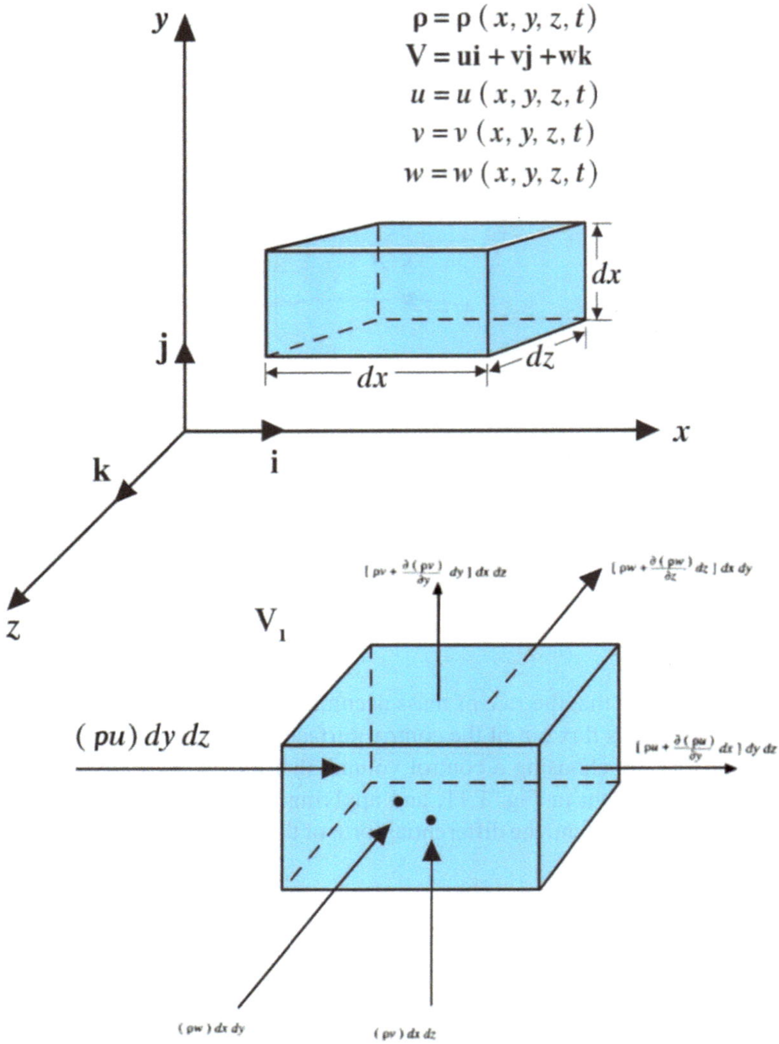

Fig. 1.11 Schematic of a moving control volume following the fluid, illustrating the principle of continuity

Consider a moving control volume (Fig. 1.12). The governing form of the momentum equation for a compressible Newtonian fluid, neglecting volumetric viscosity, is represented in the conservative form of the Navier–Stokes equations:

$$\frac{\partial(\rho u)}{\partial t}+\frac{\partial\left(\rho u^2\right)}{\partial x}+\frac{\partial(\rho uv)}{\partial y}+\frac{\partial(\rho uw)}{\partial z}=-\frac{\partial p}{\partial x}+\mu\left(\frac{\partial^2 u}{\partial x^2}+\frac{\partial^2 u}{\partial y^2}+\frac{\partial^2 u}{\partial z^2}\right)+\rho f_x \tag{1.5a}$$

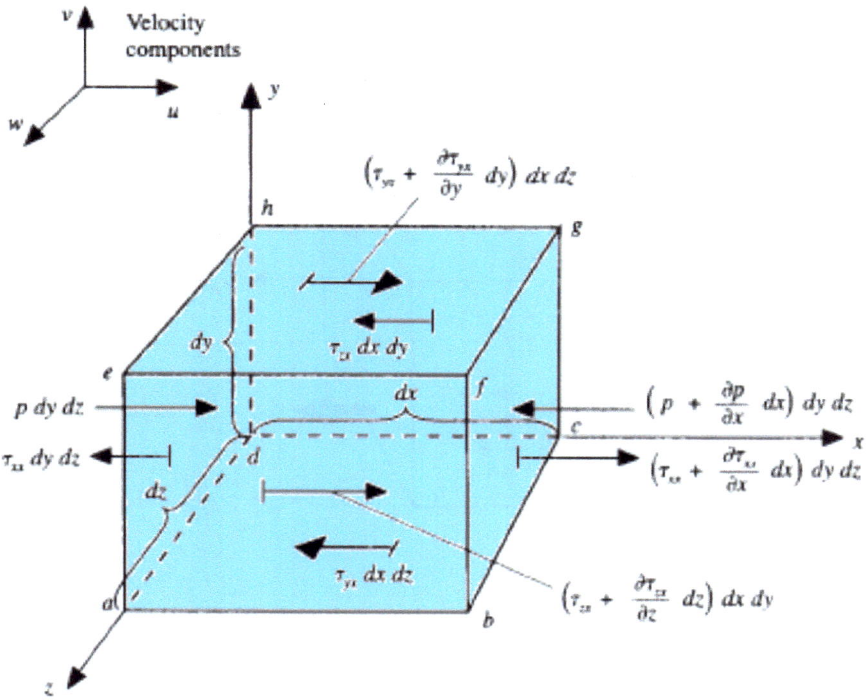

Fig. 1.12 Schematic of a moving control volume following the fluid, illustrating the principle of momentum conservation

$$\frac{\partial(\rho v)}{\partial t} + \frac{\partial(\rho u v)}{\partial x} + \frac{\partial\left(\rho v^2\right)}{\partial y} + \frac{\partial(\rho v w)}{\partial z} = -\frac{\partial p}{\partial y} + \mu\left(\frac{\partial^2 v}{\partial x^2} + \frac{\partial^2 v}{\partial y^2} + \frac{\partial^2 v}{\partial z^2}\right) + \rho f_y \quad (1.5b)$$

$$\frac{\partial(\rho w)}{\partial t} + \frac{\partial(\rho u w)}{\partial x} + \frac{\partial(\rho v w)}{\partial y} + \frac{\partial\left(\rho w^2\right)}{\partial z} = -\frac{\partial p}{\partial z} + \mu\left(\frac{\partial^2 w}{\partial x^2} + \frac{\partial^2 w}{\partial y^2} + \frac{\partial^2 w}{\partial z^2}\right) + \rho f_z \quad (1.5c)$$

where uv and w are the velocity components in the xy and z directions respectively, p is the pressure, μ is the dynamic viscosity, and $f_x f_y andf_z$ representthe body force components.

1.11.3 Principle of Energy Conservation

The energy conservation principle is essentially the first law of thermodynamics, applied to a fluid element that moves with the flow (Fig. 1.13). It states that the rate

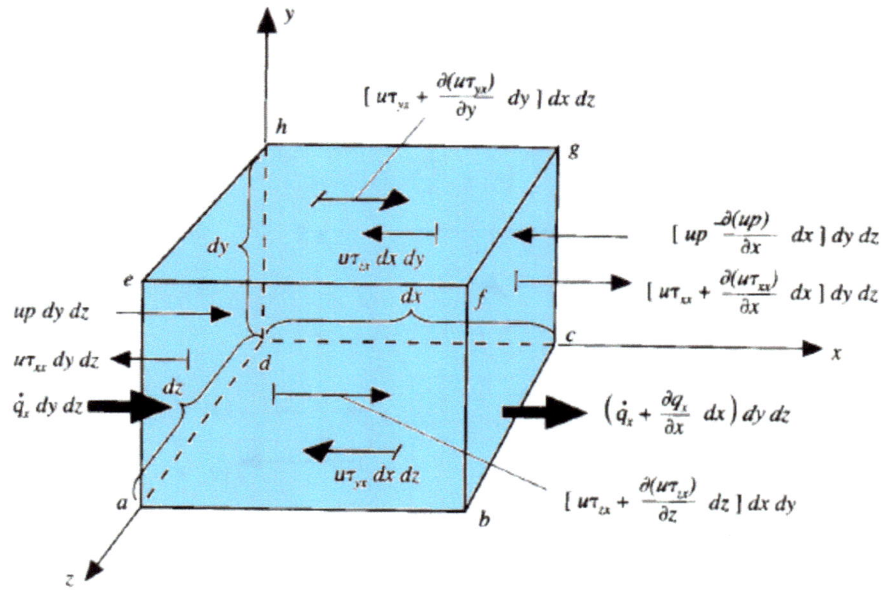

Fig. 1.13 Schematic of a moving control volume following the fluid, illustrating the principle of energy conservation

of change of total energy within the control volume is equal to the sum of the net heat added and the net work done on the control volume.

The total rate of heat addition B, accounting for internal heat generation $\dot{q}$ and conduction losses, is expressed as:

$$B = \left[\rho \dot{q} - \left(\frac{\partial \dot{q}_x}{\partial x} + \frac{\partial \dot{q}_y}{\partial y} + \frac{\partial \dot{q}_z}{\partial z} \right) \right] dx\, dy\, dz. \tag{1.6}$$

The net rate of work C done on the control volume due to pressure forces and viscous stresses is given by:

$$C = \left[\frac{\partial (up)}{\partial x} + \frac{\partial (vp)}{\partial y} + \frac{\partial (wp)}{\partial z} + \sum_{i,j} \frac{\partial \left(u_i \tau_{ij} \right)}{\partial x_j} \right] dx\, dy\, dz + \rho \boldsymbol{f} \cdot \boldsymbol{V}\, dx\, dy\, dz. \tag{1.7}$$

where

- $\dot{q}$ is the internal heat generation per unit mass,
- τ_{ij} is the viscous stress tensor components,
- $\boldsymbol{f}$ is the body force vector, and
- $\boldsymbol{V}$ is the velocity vector.

1.12 Turbulence

All fluid flows encountered in nature and engineering applications become unstable beyond a certain threshold characterized by the Reynolds number, a dimensionless parameter that quantifies the ratio of inertial to viscous forces. It is used as a critical indicator to distinguish between laminar, transitional, and turbulent flow regimes. The Reynolds number is defined as:

$$\mathrm{Re} = \frac{\rho v L}{\mu} \tag{1.8}$$

where

- ρ is the fluid density $\left[\mathrm{kg/m^3}\right]$,
- v is the characteristic (mean) velocity of the flow [m/s],
- L is a characteristic length scale [m], and
- μ is the dynamic viscosity of the fluid [Pa $\cdots$].

Flows with low Reynolds numbers tend to be laminar, characterized by smooth and orderly fluid motion. Conversely, high Reynolds numbers indicate turbulent flow, exhibiting chaotic fluctuations and eddies spanning a wide range of scales. The transitional regime lies between these two extremes, where flow begins to destabilize and shows a mix of laminar and turbulent features. Figure 1.14 illustrates the three principal regimes: laminar, transitional, and turbulent flows.

1.13 Boundary-Layer Flow Over Solid Surfaces

When a fluid flows over a solid boundary (e.g., a flat plate), a thin region forms adjacent to the surface where viscous effects dominate. This region is referred to as the boundary layer, and its structure varies depending on whether the flow is laminar or turbulent. Figure 1.15 shows the typical development of such a boundary layer.

In the laminar regime, fluid particles move in well-defined parallel layers. However, beyond a critical Reynolds number, the flow undergoes a transition and becomes turbulent, characterized by rotational structures and random fluctuations.

The turbulent boundary layer is stratified into several distinct subregions:

- **Viscous sublayer**: A thin region adjacent to the wall where the flow remains essentially laminar, and viscous shear dominates. Fluid motion is smooth and linear with respect to the wall.
- **Buffer layer**: An intermediate zone between the viscous sublayer and the outer turbulent region, where neither viscous nor turbulent effects are negligible. This region is highly complex and transitional in behavior.

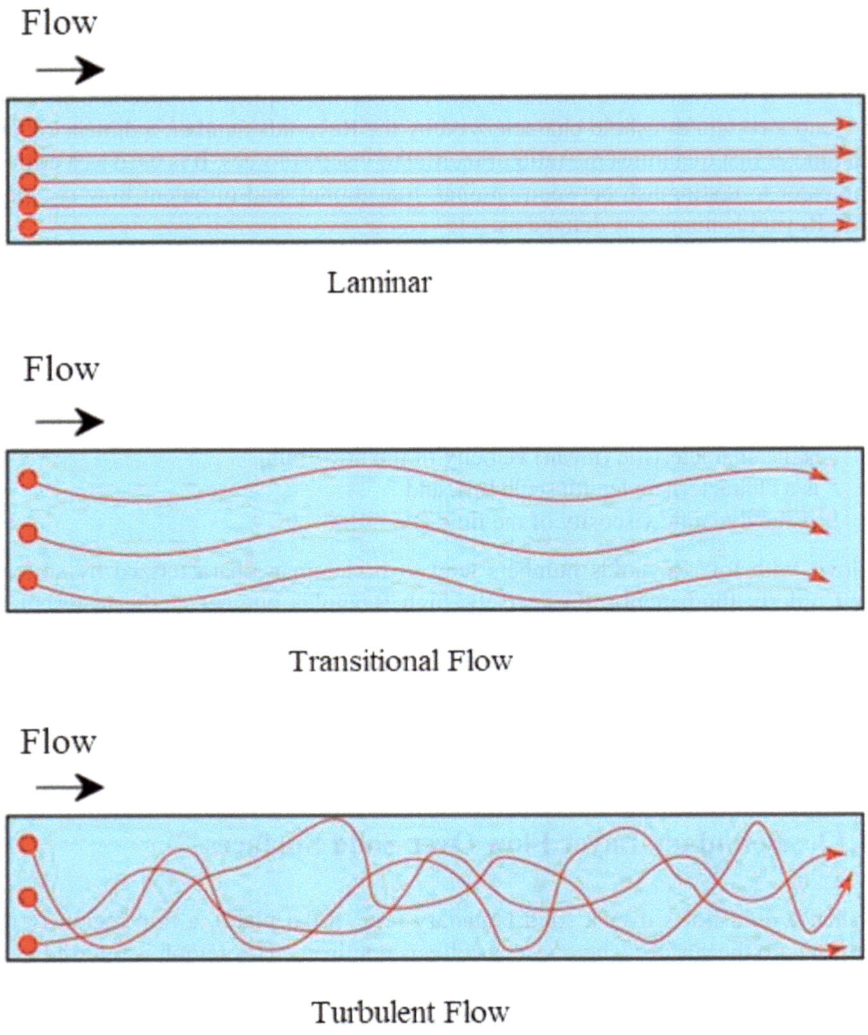

Fig. 1.14 Schematic illustration of laminar, transitional, and turbulent flow regimes

- **Log-law or overlap layer**: This zone connects the buffer and outer turbulent regions. The velocity distribution follows the logarithmic law of the wall and is sensitive to Reynolds number variations.

The precise structure of these sublayers is critical for accurate turbulence modeling, especially in wall-bounded flows where near-wall treatment strongly influences drag prediction and heat transfer.

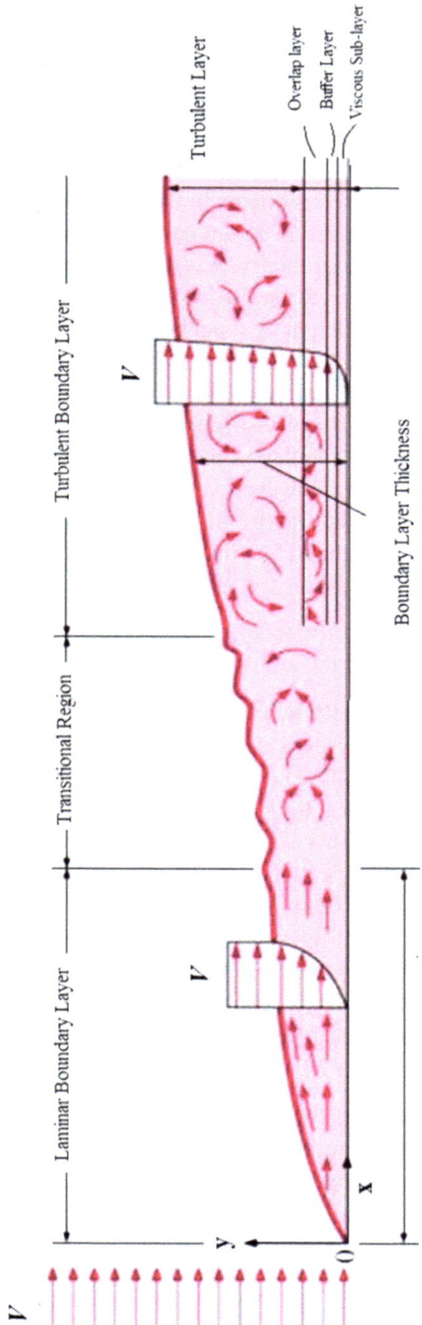

Fig. 1.15 Laminar boundary layer, transitional region, and turbulent boundary layer of a fluid flow over a surface

1.14 Atmospheric Boundary Layer

To analyze the behavior of fluid flow around buildings exposed to atmospheric free-stream conditions, it is essential to characterize the boundary layer of the external flow and determine the flow regime. The atmospheric boundary layer (ABL), like other boundary layers, possesses distinct characteristics and physical scaling properties. In many practical cases, the Earth's surface can be assumed to be aerodynamically rough, allowing the neglect of viscous effects in the outer layer.

The turbulent boundary layer over an aerodynamically rough surface is primarily governed by the surface roughness geometry. In regions away from local obstructions, the flow behavior is typically characterized using the aerodynamic roughness length z_0, which represents the average height of roughness elements.

Assuming an analogy between the atmospheric boundary layer and a fully rough turbulent boundary layer over a flat plate, it is possible to simulate the atmospheric layer in wind tunnel experiments using appropriate physical modeling. When constructing a scaled model, the roughness parameter z_0 must be reduced proportionally to the geometric scaling. However, care must be taken to ensure that the Reynolds number does not fall below the threshold at which the surface can be considered aerodynamically rough. Achieving a fully developed boundary layer in such setups typically requires a sufficiently long fetch in the wind tunnel.

It is important to note two major differences between a canonical turbulent boundary layer over a smooth flat plate and the atmospheric boundary layer:

1. **Wind direction variability with height**: In the ABL, the primary flow direction often changes with elevation due to Coriolis effects and thermal stratification. This directional shear can significantly influence the dispersion of pollutants and momentum transfer. In contrast, the flow direction in a flat plate boundary layer remains essentially uniform. This effect is difficult to replicate accurately in wind tunnel testing and is often neglected in preliminary studies.
2. **Buoyancy effects**: Resulting from temperature gradients, buoyancy can either suppress or enhance turbulent intensity. Such buoyancy-induced modifications to turbulence strongly affect the vertical transport of heat, momentum, and scalar quantities such as pollutants. These limitations also affect the validity of using simplified ABL simulations for structural loading assessments involving dynamic and static forces.

For accurate simulation, the ABL must be established within the shortest feasible fetch while maintaining appropriate roughness characteristics based on the geometric scaling of the model. A realistic boundary layer is achieved only when the flow is allowed to develop naturally over a surface whose roughness elements are scaled in accordance with the physical model.

The structure of wind in the free atmosphere is highly complex. To fully describe the characteristics of atmospheric flow, the following parameters are typically required:

1. Vertical profile of the mean wind speed;
2. Turbulence intensity of each velocity component as a function of height;
3. Integral length scales of turbulence;
4. Energy-containing eddy scales (turbulence energy scales);
5. Cross-correlations between turbulent components, particularly Reynolds stresses;
6. Turbulence spectra.

Turbulence spectra and turbulence length scales provide two complementary descriptions of turbulent flow. Depending on the application, one may be preferred over the other. In practical atmospheric studies, spectral analysis is often favored due to its relative ease of measurement and its direct connection to energy distribution across scales.

1.14.1 Mean Wind Velocity Profile with Height

The wind speed in the atmospheric boundary layer varies with height and is not constant. Near the ground surface, wind velocity is essentially zero due to the no-slip condition and increases with height according to the empirical power-law profile given as:

$$\frac{V_h}{V_H} = \left(\frac{h}{H}\right)^{\alpha} \tag{1.9}$$

Here, V_h is the wind speed at height h, V_H is the wind speed at reference height H, and α is the surface roughness exponent that depends on the terrain type. Empirical values of α typically fall within the following ranges:

- Rural areas: $0.10 < \alpha < 0.30$
- Suburban zones: $0.30 < \alpha < 0.50$
- Urban and dense city centers: $0.50 < \alpha < 1.0$

This profile is widely used in boundary layer meteorology and wind engineering to approximate velocity gradients near the ground.

1.14.2 Turbulence Intensity

A key parameter in the characterization of turbulent flows is the instantaneous velocity of the fluid, which is inherently time-dependent and stochastic in turbulent regimes. The instantaneous velocity components in the Cartesian directions are represented by:

$$U(t) = \overline{U} + u'(t) \tag{1.10}$$

$$V(t) = \overline{V} + v'(t) \tag{1.11}$$

$$W(t) = \overline{W} + w'(t) \tag{1.12}$$

where $\overline{UV}$ and $\overline{W}$ denote the time-averaged (mean) velocity components, and $u'(t)$, $v'(t)$, and $w'(t)$ represent the fluctuating components (also called turbulent fluctuations). These fluctuations are essential in understanding the dynamics of turbulent flow, especially in computing Reynolds stresses and turbulent kinetic energy.

The turbulence intensity (TI) for each velocity component is defined as the ratio of the root-mean-square (RMS) of the fluctuating velocity to the mean velocity in the corresponding direction:

$$T_u = \frac{u'}{U_{\text{mean}}} \times 100\% \tag{1.13}$$

$$T_v = \frac{v'}{V_{\text{mean}}} \times 100\% \tag{1.14}$$

$$T_w = \frac{w'}{W_{\text{mean}}} \times 100\% \tag{1.15}$$

These nondimensional measures provide insight into the level of unsteadiness and mixing within each directional component of the flow. Turbulence intensities can vary significantly depending on the flow regime, surface roughness, and Reynolds number. In fully turbulent atmospheric boundary layers, these values are critical for pollutant dispersion, wind loading on structures, and microclimate assessment.

1.14.3 Turbulence Energy Spectrum

In turbulent flows, fluid motion is characterized by rotational structures of various scales, known as eddies (Fig. 1.16). Large eddies serve as the primary carriers of energy, transferring it through an energy cascade process to smaller eddies, which act

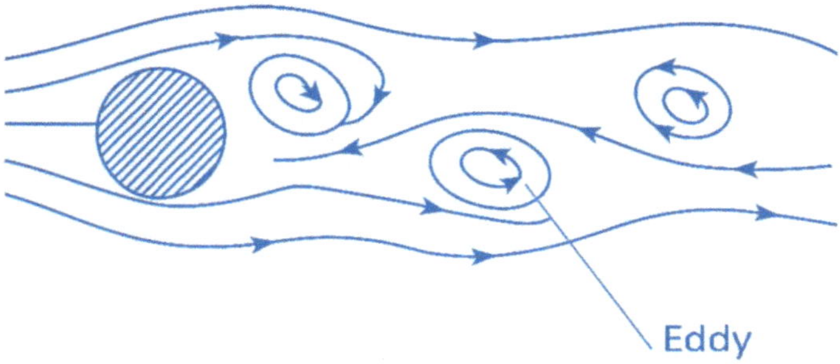

Fig. 1.16 Generation of vortices due to fluid flow past a cylindrical object

as energy dissipators. Conceptually, if each eddy is associated with a characteristic wavenumber κ, the transfer of energy progresses from low wavenumbers (large eddies) to high wavenumbers (small eddies). Thus, small-scale eddies absorb the energy transferred from larger-scale motions.

Each eddy consists of smaller embedded eddies in a hierarchical structure that ultimately culminates in molecular motion—defining the smallest eddy, which corresponds to the Kolmogorov scale. While large eddies are generally inhomogeneous, small eddies are assumed to be homogeneous and isotropic. The turbulent flow field is therefore composed of a wide range of eddy sizes, as depicted in Fig. 1.17.

The fragmentation of larger eddies into smaller ones results in the cascade of kinetic energy, a process widely referred to as the energy cascade. At sufficiently small scales, kinetic energy is dissipated as heat due to the action of fluid viscosity.

At high Reynolds numbers, the energy spectrum $E(\kappa)$ follows the Kolmogorov $-5/3$ law, where the energy content is distributed according to the power law:

$$E(\kappa) \propto \varepsilon^{2/3}\kappa^{-5/3} \tag{1.16}$$

Here

- κ is the wavenumber (inverse of length scale), and
- ε is the mean rate of turbulent kinetic energy dissipation per unit mass.

This relationship is derived through dimensional analysis and is valid in the inertial subrange, where the influence of viscosity is negligible but the energy transfer remains active.

More generally, the energy spectrum is expressed as:

$$E = E(\varepsilon\kappa) \tag{1.17}$$

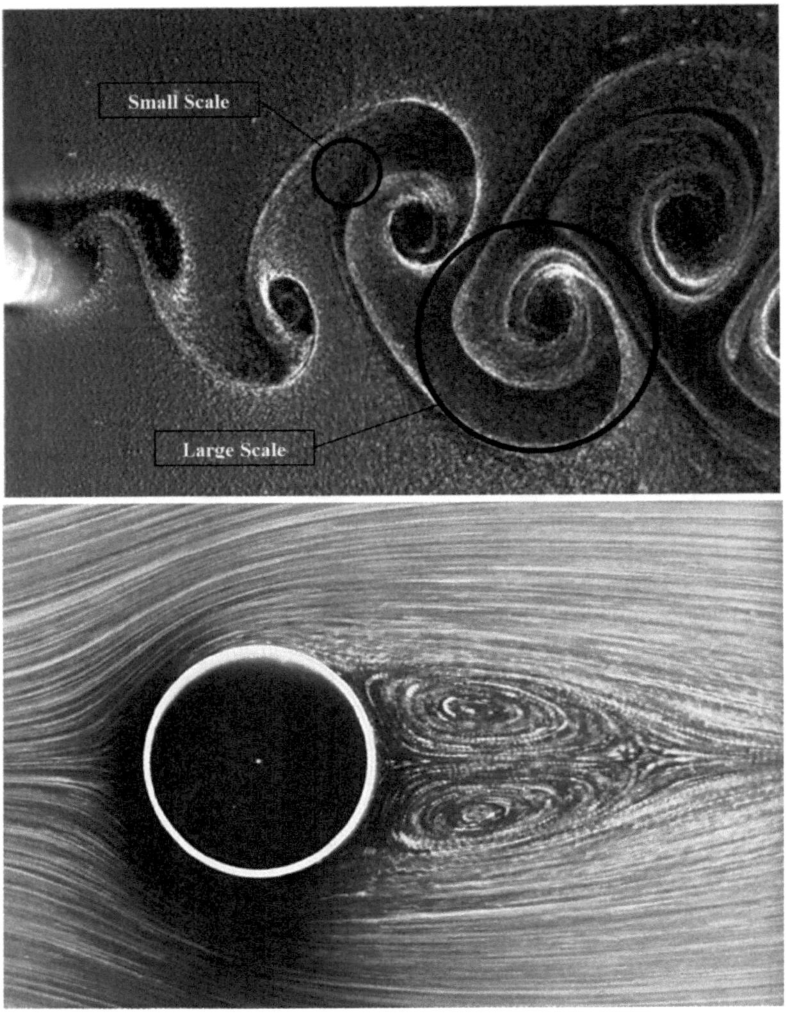

Fig. 1.17 Representation of eddies at multiple scales generated by fluid interaction with a solid body

1.14.3.1 Classification of Turbulent Eddies

Turbulent eddies can be classified into three major categories based on their wavenumber:

(a) *Large Eddies—Integral Scale Region*

At low wavenumbers, the energy spectrum is highly non-uniform. The integral length scale L is comparable to the characteristic geometric scale of the physical system.

The associated integral time scale τ and length scale L can be estimated as:

$$L = \frac{k^{3/2}}{\varepsilon} \tag{1.18}$$

$$\tau = \frac{k}{\varepsilon} \tag{1.19}$$

where

- k is the turbulent kinetic energy per unit mass, and
- ε is the dissipation rate.

(b) *Intermediate Scales—Taylor Microscale Region*

This regime corresponds to intermediate wavenumbers and is commonly referred to as the *Taylor microscale region.* The eddies in this range follow the Kolmogorov energy spectrum and exhibit nearly universal behavior across different turbulent flows:

$$E(\kappa) = c_k \, \varepsilon^{2/3} \, \kappa^{-5/3} \tag{1.20}$$

where

- $E(\kappa)$ is the turbulent kinetic energy spectrum as a function of wavenumber κ,
- ε is the turbulent kinetic energy dissipation rate, and
- $c_k \approx \frac{1}{3}$ is the Kolmogorov constant.

The *Taylor microscale* λ represents a characteristic length scale at which viscous effects begin to influence the turbulent energy cascade but are not yet dominant. It is estimated by the relation:

$$\lambda = \left(\frac{10 \, \nu \, k}{\varepsilon} \right)^{1/2} \tag{1.21}$$

where

- ν is the kinematic viscosity of the fluid,
- k is the turbulent kinetic energy, and
- ε is the energy dissipation rate.

(c) *Smallest Eddies—Kolmogorov Dissipation Range*

At very high wavenumbers, the smallest scales of turbulence are governed by viscous effects. This regime is termed the Kolmogorov scale, where energy is rapidly dissipated. The energy spectrum decreases exponentially with increasing κ, and eddies in

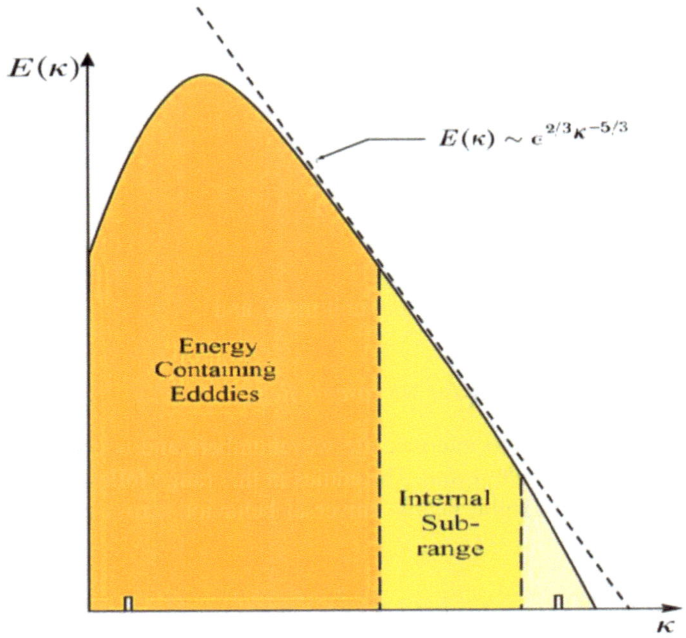

Fig. 1.18 Energy spectrum of a turbulent flow as a function of wavenumber

this range possess minimal kinetic energy. The Kolmogorov length scale η is defined as:

$$\eta = \left(\frac{\nu^3}{\varepsilon}\right)^{1/4} \tag{1.22}$$

Figure 1.18 illustrates the turbulence energy spectrum $E(\kappa)$ over the full wavenumber range, highlighting the three major regions described above: the energy-containing range, inertial subrange, and dissipation range.

1.15 Boundary Conditions

A boundary layer is defined as a region of fluid that is influenced by heat, momentum, or mass transfer originating from an adjacent surface, which may be either stationary or in motion. A classic example is the layer of fluid adjacent to a heat source, such as a radiator. The fluid in this near-wall region absorbs thermal energy, thereby forming a temperature gradient, which leads to density variations and can initiate convective flow.

In fluid dynamics, phenomena such as drag forces and flow separation are closely linked to changes in the shape and thickness of the velocity boundary layer around a moving object. Boundary conditions specify the thermal and flow parameters along the computational domain's boundaries. Consequently, the accurate prescription of boundary conditions is a critical step in the numerical simulation of fluid flow.

In computational fluid dynamics (CFD) software, boundary conditions generally fall into the following categories:

- **Inlet and Outlet Boundaries**: These include velocity inlets, pressure inlets, mass flow inlets, intake vents, inlet fans, far-field pressure boundaries, outflow boundaries, outlet vents, and outlet fans.
- **Wall and Symmetry Boundaries**: These include solid walls, periodic boundaries, symmetry boundaries, and axisymmetric conditions.
- **Internal Zones**: These refer to regions within the computational domain that consist of fluid or solid material.
- **Internal Surfaces**: These include boundary conditions for components such as fans, radiators, porous jumps, solid walls, and interior interfaces.

Internal surface boundary conditions are applied to the faces of interior computational cells. Such surfaces possess a finite thickness, and flow parameters on either side of the face may exhibit significant differences. These types of boundary conditions are typically used to impose physical characteristics of devices like fans, thin porous media, and radiators.

An exception is the interior boundary condition, which does not represent a physical obstruction to fluid flow. In essence, an interior boundary simply demarcates regions within the domain without acting as a physical boundary.

1.15.1 Inlet and Outlet Boundary Conditions

In CFD software such as ANSYS Fluent, a wide range of boundary condition types are available to define the flow characteristics at domain inlets and outlets. This section aims to enhance the user's understanding of each type by outlining their specific applications and characteristics. Atmospheric boundary layer for analyzing wind flow around a building as case study is illustrated in Fig. 1.19.

General Use of Boundary Conditions

In most CFD applications, approximately ten different boundary condition types are available to define inlet and outlet flows. These include:

- Pressure inlet,
- Velocity inlet,
- Mass flow inlet,
- Pressure outlet,
- Outflow

Fig. 1.19 Atmospheric boundary layer for analyzing wind flow around a building

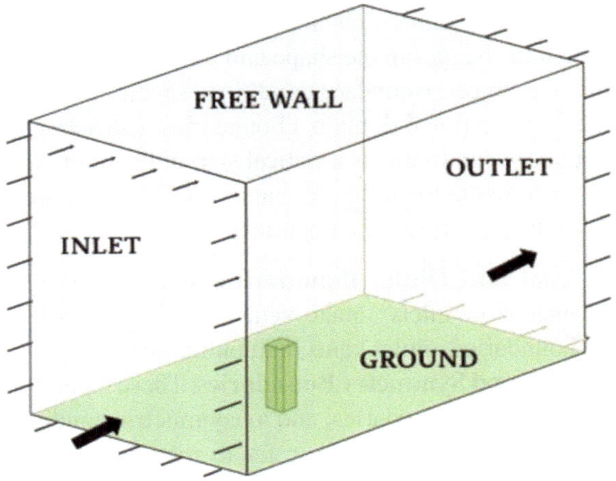

- Far-field pressure,
- Inlet vent,
- Outlet vent,
- Inlet fan, and
- Outlet fan.

Each of these boundary condition types serves a distinct modeling purpose. A summary of their applicability is typically presented in tables, such as Table 1.1, which outlines usage guidelines for each type.

1.15.2 Application in Building Boundary Layer Modeling

Figures 1.20 and 1.21 illustrate two practical configurations for modeling airflow in and around buildings using boundary layer concepts.

Case 1: Isolated Indoor Domain (No External Wind Exposure)

In the first scenario (Fig. 1.20a), the building is assumed to be isolated from the external environment, with no direct wind flow impinging on the structure. Airflow enters and exits the interior domain exclusively through openings such as windows or vents, which serve as the inlet and outlet boundaries, respectively. For each inlet or outlet, the following flow parameters must be specified:

- Air velocity,
- Static pressure,
- Temperature, and
- Volumetric or mass flow rate.

Table 1.1 Application of boundary conditions in CFD software

Boundary condition	Definition
Velocity inlet	Used to prescribe the velocity vector and scalar properties of the flow at inlet boundaries
Pressure inlet	Applied to specify the total (stagnation) pressure and other flow quantities at the inlet
Mass flow inlet	Typically used in compressible flows to define the inlet mass flow rate. Not required for incompressible flows, since a specified velocity inlet ensures a constant mass flow rate due to constant density
Pressure outlet	Used to set the static pressure and other scalar variables at the outlet. Compared to the outlet flow boundary, this condition often leads to improved convergence, especially in the presence of backflow during iterations
Pressure far-field	Applied only in simulations involving external compressible flows. It defines free-stream Mach number and far-field static conditions
Outflow	Suitable for cases with fully developed outflow (e.g., in long internal ducts). Not recommended for compressible flows. Used when detailed information at the outlet is not available
Vent inlet	Models an inlet vent with specified pressure loss coefficient, flow direction, and ambient total pressure and temperature
Fan inlet	Represents an inlet fan located upstream of the physical domain, characterized by a specified pressure jump, flow direction, and ambient total pressure and temperature
Vent outlet	Models an outlet vent with given pressure loss coefficient and specified static pressure and temperature
Fan outlet	Represents an outlet fan that imposes a pressure jump, with known flow direction and ambient total pressure and temperature

Additionally, internal openings between zones within the building are assigned the interior boundary condition type (also shown in Fig. 1.20a). In Fluent, interior boundaries are treated as transparent interfaces that allow mesh continuity and flow passage without obstruction. These surfaces can be meshed and used in the simulation as part of the computational domain.

Next, solar radiation incident on various building surfaces—both transparent (glazed) and opaque—should be defined. After specifying the boundary layer configuration, a computational grid (mesh) is generated for the interior volume (Fig. 1.20b), and the CFD simulation is performed. Postprocessing yields results such as wind velocity vectors, temperature fields, and pressure distributions in both numerical and graphical formats (Fig. 1.20c).

Case 2: Wind-Driven Outdoor Flow Over Building

In the second scenario, the building is exposed to wind blowing from a distant source (Fig. 1.21a). Here, the domain includes the surrounding outdoor region, which is typically represented by a rectangular block of air. One face of this block is defined as a velocity inlet, while the opposite face is defined as a pressure outlet.

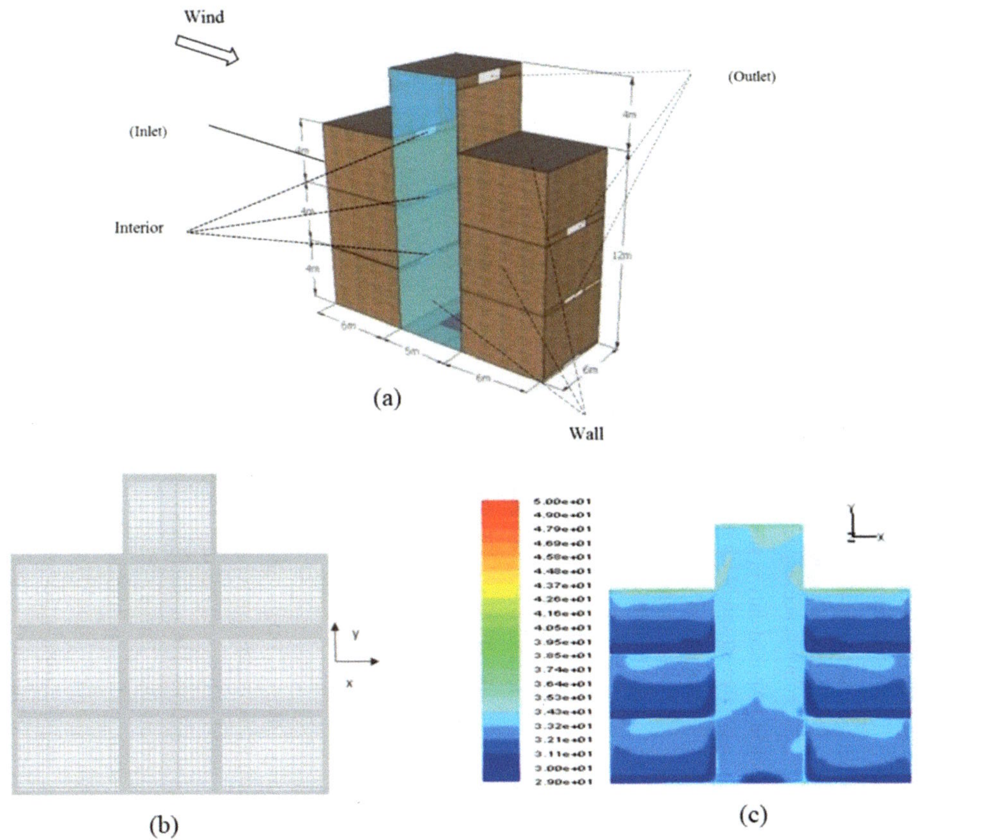

Fig. 1.20 An example atrium and associated boundary layer (**a**), computational mesh (**b**), and simulation output using ANSYS Fluent (**c**)

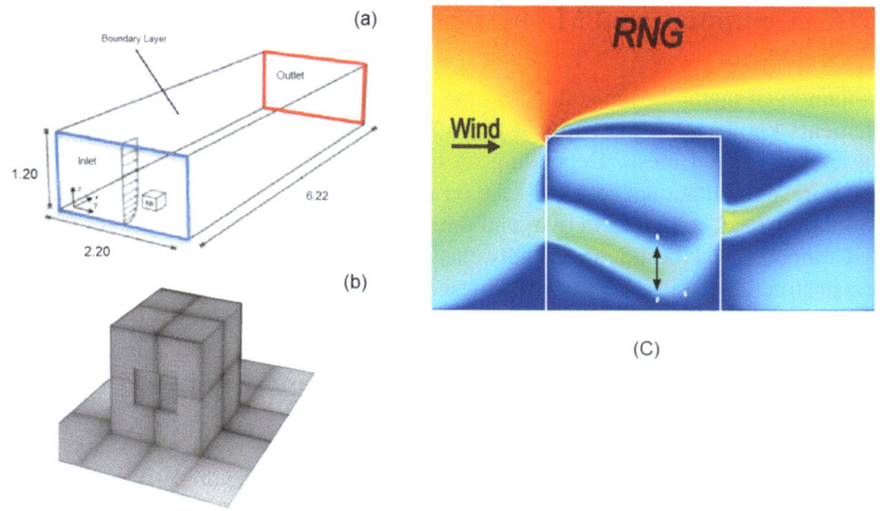

Fig. 1.21 CFD simulation of airflow through a room: atmospheric boundary layer around a room (**a**), computational mesh of the surrounding domain (**b**), and simulated wind velocity field inside and around the room (**c**)

Key inlet parameters must be defined:

- Inflow wind velocity,
- Static pressure, and
- Inflow temperature.

Openings such as windows and doors are again assigned the interior boundary condition type, allowing for flow interaction between indoor and outdoor spaces. Solar radiation conditions—both magnitude and direction—can be specified to account for thermal loading on building surfaces. This solar radiation modeling is applicable to both cases and can significantly enhance the physical accuracy of thermal-flow simulations.

Following the boundary definition, the computational mesh is generated for the exterior domain (Fig. 1.21b), including building surfaces and interior spaces. Once the mesh is finalized, the CFD solver can be executed, and the resulting airflow velocity, temperature, and pressure fields can be visualized or extracted numerically (Fig. 1.21c).

1.16 Computational Mesh

In numerical methods, the solution domain is first discretized into numerous small elements known as computational cells. The collection of these cells constitutes the computational mesh. Depending on the geometric complexity and the solver structure, computational cells can take on various shapes. Figure 1.22 illustrates six commonly used cell types in two- and three-dimensional mesh generation.

Computational meshes are broadly classified into two main categories: structured and unstructured grids. In structured meshes, adjacent cells are indexed in a matrix-like fashion, resulting in a regular and organized grid structure (Fig. 1.23). The cells in this type of mesh are typically quadrilateral in two dimensions and hexahedral in three dimensions. However, due to their geometric constraints, structured grids are not suitable for many complex geometries.

For complex geometries, unstructured meshes are employed (Figs. 1.24 and 1.25). These meshes are generated in an irregular—often quasi-random—fashion using intelligent meshing algorithms. The predominant cell types in unstructured grids are triangles in 2D and tetrahedra in 3D, though pyramidal and prismatic cells may also be used in specific applications. Moreover, cell types typically used in structured

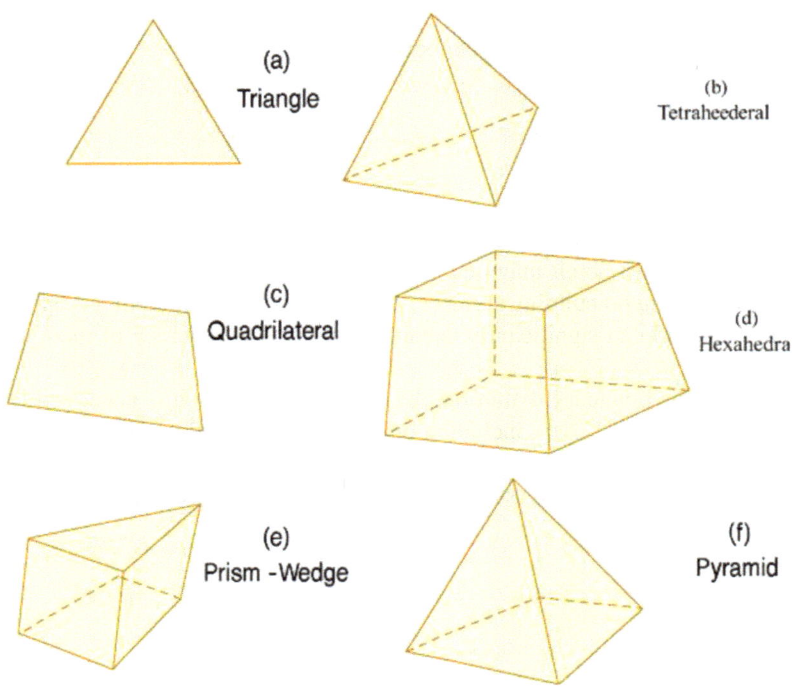

Fig. 1.22 Common types of cells used in CFD simulations

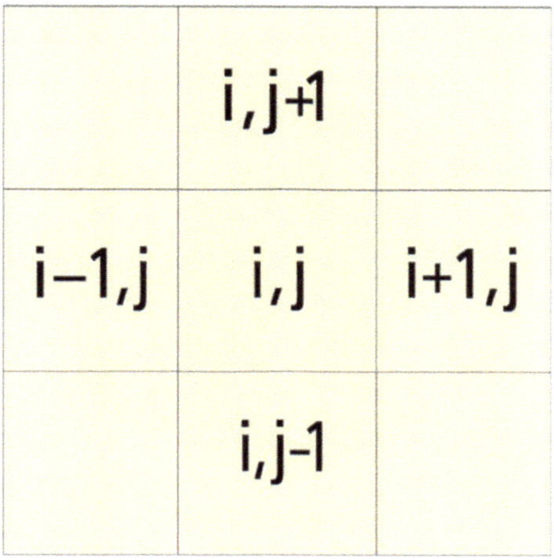

Fig. 1.23 Indexing of neighboring cells in a structured grid

grids (e.g., quadrilaterals and hexahedra) can also be utilized in unstructured meshes, though they are less common compared to the primary unstructured elements.

Examples of structured and unstructured mesh configurations are shown in Fig. 1.26. It is worth noting that commercial CFD solvers such as ANSYS FLUENT and ANSYS CFX are inherently designed for unstructured grid solvers. Nonetheless, structured meshes can still be implemented in these solvers by internally converting the indexing scheme into an unstructured format.

The discretization of the computational domain into very small cells enables the assumption that variations in the governing variables over each individual cell are linear—or even negligible. This simplification facilitates the linearization and

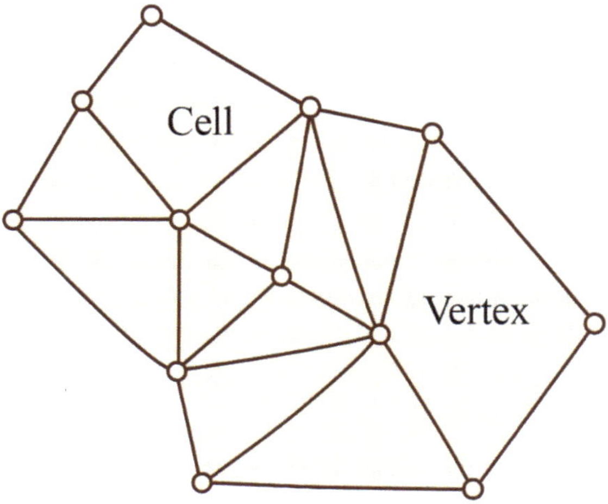

Fig. 1.24 General view of an unstructured mesh

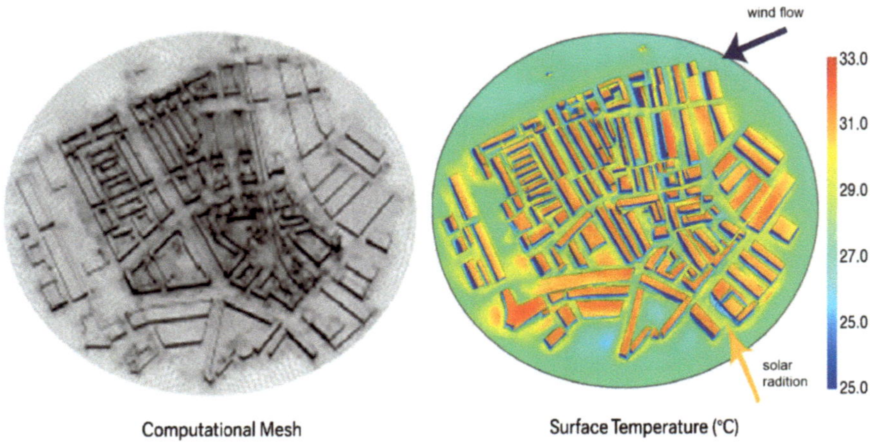

Fig. 1.25 Computational mesh over an urban district and the simulated surface temperature of buildings

discretization of the Navier–Stokes equations over each control volume. By integrating and establishing inter-cell relationships across the entire domain, the solver reconstructs an approximate representation of the fluid flow field.

As the number of cells increases and the individual cell size decreases, assumptions regarding linearity or constancy of flow variables within each cell become more valid, enhancing the fidelity of the numerical results. This methodology constitutes the foundation of CFD, a branch of fluid mechanics where the governing equations are solved using numerical techniques. Modern CFD software packages, built upon these principles, are widely used across engineering disciplines including mechanical and aerospace engineering, hydraulics, chemical processing, and biomedical applications. A concise overview of the most widely used CFD software tools is provided in Chap. 5.

1.17 Summary

This chapter introduced the fundamentals of CFD and established its role as a core analytical tool in architecture and urban planning. It presented CFD as a complement to analytical and experimental methods, capable of simulating airflow behavior, thermal comfort, and pollutant dispersion in complex-built environments. Key fluid dynamics principles—such as control volume and differential formulations, conservation laws, turbulence, and boundary layer theory—were discussed within the context of the urban microclimate. The importance of mesh generation and flow modeling strategies was emphasized through architectural and urban-scale examples. This foundation prepares the reader for the theoretical and mathematical formulations of CFD governing equations presented in Chap. 2.

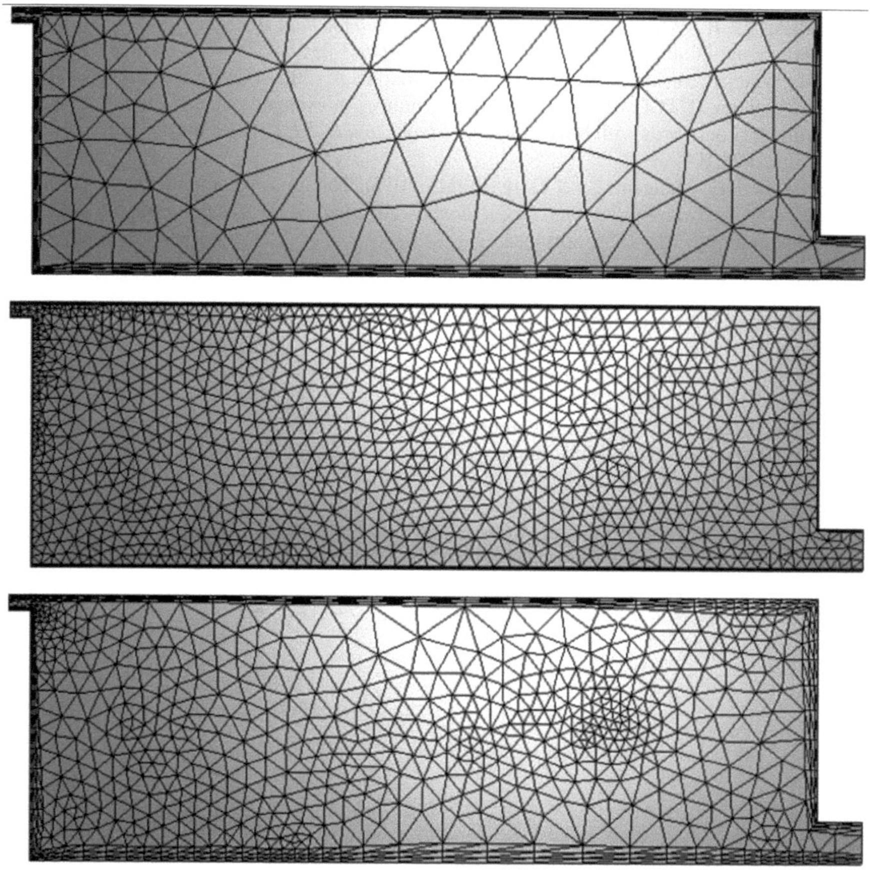

Fig. 1.26 Two-dimensional computational mesh generated by Triangle within a rectangular domain at three different resolutions: fine, medium, and coarse

References

1. Jayanti, S. (2018). *Computational fluid dynamics for engineers and scientists*. Springer.
2. Ghaib, K. (2022). *Introduction to computational fluid dynamics*. Springer Nature.
3. Cheshmehzangi, A. (2016). Multi-spatial environmental performance evaluation towards integrated urban design: A procedural approach with computational simulations. *Journal of Cleaner Production, 139*, 1085–1093. https://doi.org/10.1016/j.jclepro.2016.08.151
4. Cheshmehzangi, A., Zhu, Y., & Li, B. (2010). Integrated design approach-urban design for sustainability. In *5th International Conference on Responsive Manufacturing-Green Manufacturing (ICRM 2010)* (pp. 241–247). IET.
5. Lecheler, S. (2022). *Computational fluid dynamics*. Springer Fachmedien Wiesbaden. https://doi.org/10.1007/978-3-658-38453-1

6. Maliska, C. R. (2023). *Fundamentals of computational fluid dynamics* (Vol. 135). Springer International Publishing. https://doi.org/10.1007/978-3-031-18235-8
7. Mahdavinejad, M., Shaeri, J., Nezami, A., & Goharian, A. (2024). Comparing universal thermal climate index (UTCI) with selected thermal indices to evaluate outdoor thermal comfort in traditional courtyards with BWh climate. *Urban Climate, 54*, 101839. https://doi.org/10.1016/j.uclim.2024.101839
8. Moukalled, F., Mangani, L., & Darwish, M. (2015). The finite volume method. In *The finite volume method in computational fluid dynamics: An advanced introduction with OpenFOAM® and Matlab* (pp. 103–135). Springer.
9. Shaeri, J., Mahdavinejad, M., Vakilinejad, R., Bazazzadeh, H., & Monfared, M. (2023). Effects of sea-breeze natural ventilation on thermal comfort in low-rise buildings with diverse atrium roof shapes in BWh regions. *Case Studies in Thermal Engineering, 41*, 102638.
10. Shaeri, J., Aflaki, A., Yaghoubi, M., & Janalizadeh, H. (2018). Investigation of passive design strategies in a traditional urban neighborhood: A case study. *Urban Climate, 26*, 31–50. https://doi.org/10.1016/j.uclim.2018.08.003
11. Shaeri, J., Mahdavinejad, M., & Pourghasemian, M. H. (2022). A new design to create natural ventilation in buildings: Wind chimney. *Journal of Building Engineering, 59*, 105041. https://doi.org/10.1016/j.jobe.2022.105041
12. Shaeri, J., & Mahdavinejad, M. (2022). Prediction indoor thermal comfort in traditional houses of Shiraz with PMV/PPD model. *International Journal of Ambient Energy, 43*(1), 8316–8334.

Chapter 2
CFD Theory and Governing Equations at Building and Urban Scales

Abstract This chapter presents the mathematical formulation of two-dimensional, steady-state, and turbulent airflow and its coupling with multi-mode heat transfer over building envelopes. The governing equations for fluid motion are derived from the fundamental conservation laws of mass, momentum, and energy, incorporating the RNG K–ε turbulence model to represent Reynolds-averaged turbulent fluctuations. These equations are expressed in generalized curvilinear coordinates, allowing simulation over complex architectural geometries such as sloped or non-orthogonal roof structures. Assumptions of Newtonian fluid behavior, constant thermophysical properties, and negligible body forces are employed to simplify the modeling framework. The mathematical formulation includes coordinate transformations using covariant and contravariant basis vectors, as well as Jacobian and metric tensors, facilitating control volume construction for arbitrary geometries frequently encountered in architectural and urban contexts. The heat transfer model integrates three distinct mechanisms: conduction through solid roof layers, forced convection induced by wind–surface interactions, and radiative exchange with both solar and sky sources. Each mechanism is incorporated through dedicated source terms within the energy equation, providing a physically grounded representation of building-scale thermal processes. Designed with architects and urban planners in mind, this formulation equips the reader with the theoretical tools necessary to simulate the interaction between wind and building surfaces under realistic environmental conditions. The models developed in this chapter form the computational foundation for the numerical methods introduced in Chap. 3.

Keywords Turbulent urban airflow modeling · RNG k–ε turbulence model · Generalized curvilinear coordinates · Coupled heat and flow simulation · Dimensionless governing equations

J. Shaeri and A. Cheshmehzangi, *Applied CFD*, Urban Sustainability,
https://doi.org/10.1007/978-981-95-8505-2_2

2.1 Introduction

This chapter presents the theoretical framework and governing equations required for modeling two-dimensional turbulent airflow and heat transfer in the built environment. The analysis is divided into two main components: fluid dynamics and thermal transport. Atmospheric wind around buildings is inherently turbulent and dominated by boundary layer behavior. Therefore, to accurately predict wind flow patterns around structures, the governing equations must incorporate turbulence modeling techniques.

Turbulence is accounted for using the RNG k–ε model, which enables the statistical representation of turbulent fluctuations in terms of mean flow variables. The governing equations are formulated under the following simplifying assumptions:

- Two-dimensional, steady-state, turbulent flow,
- Newtonian, incompressible fluid,
- Constant thermophysical properties throughout the flow domain, and
- Negligible body forces.

In the second part of this chapter, the energy transport mechanisms are analyzed, including conduction through the roof layer, forced convection due to wind-driven flow, and radiative heat exchange from solar and sky sources. These mechanisms are formulated through appropriate heat transfer equations and are subsequently integrated into the overall computational framework. The inclusion of both convective and radiative effects provides a more comprehensive understanding of building energy behavior under varying atmospheric conditions.

2.2 Governing Equations for Two-Dimensional Turbulent Flow

Turbulent flow is inherently unsteady, irregular, and chaotic in nature. This flow regime is characterized by the presence of eddies—vortical fluid structures—with a wide range of length scales. The size of these eddies depends on the geometric features of the flow domain, such as boundary layer thickness and surface curvature. Small-scale eddies are particularly influenced by viscous forces, which convert kinetic energy into internal energy through shear stresses [1, 2]. The governing framework for modeling turbulent flow is still based on the Navier–Stokes equations, which describe the conservation of momentum in a viscous, incompressible fluid.

In turbulent flow, the rate of momentum diffusion is significantly enhanced. Consequently, the boundary layer grows faster compared to that in laminar flow [3]. Turbulence promotes momentum exchange in the transverse direction, which can delay or mitigate the separation of the boundary layer from solid surfaces.

The tensorial form of the continuity and momentum equations for incompressible, unsteady turbulent flow—based on the assumptions introduced earlier—is given by:

$$\frac{\partial u_i}{\partial x_i} = 0 \tag{2.1}$$

$$\rho\left(\frac{\partial u_i}{\partial t} + u_j\frac{\partial u_i}{\partial x_j}\right) = -\frac{\partial P}{\partial x_i} + \mu\nabla^2 u_i \tag{2.2}$$

where

- u_i is the velocity component in the x_i direction,
- P is the pressure,
- ρ is the fluid density,
- μ is the dynamic viscosity,
- t is time, and
- ∇^2 denotes the Laplacian operator.

Due to the stochastic nature of turbulence, all flow variables exhibit random fluctuations. Consequently, the direct numerical solution of the full Navier–Stokes equations—known as direct numerical simulation (DNS)—becomes computationally infeasible for high Reynolds number flows. This is because resolving the smallest eddies in the flow requires an exceedingly large number of grid points and time steps, resulting in a computational load that demands advanced high-performance computing (HPC) resources.

From an engineering perspective, time-averaged flow quantities are often of greater interest than instantaneous values. As such, it becomes necessary to derive governing equations for the mean flow field. To this end, each instantaneous flow variable is decomposed into a mean and a fluctuating component, expressed as:

$$\phi = \overline{\phi} + \phi' \tag{2.3}$$

where

- $\overline{\phi}$ is the time-averaged component, and
- ϕ' is the fluctuating (turbulent) component.

The mean value is defined as:

$$\overline{\phi} = \frac{1}{T}\int_0^T \phi \, dt \tag{2.4}$$

Here, T is the averaging period, assumed to be sufficiently long for statistically steady behavior. Applying this decomposition to the velocity and pressure fields yields:

$$u_i = \overline{u_i} + u_{i'} \tag{2.5}$$

$$P = \overline{P} + P' \tag{2.6}$$

By substituting Eqs. (2.5) and (2.6) into the Navier–Stokes equations and applying time averaging, the Reynolds-averaged Navier–Stokes (RANS) equations are obtained:

$$\rho\left(\frac{\partial \overline{u_i}}{\partial t} + \overline{u_j}\frac{\partial \overline{u_i}}{\partial x_j}\right) = -\frac{\partial \overline{P}}{\partial x_i} + \mu \nabla^2 \overline{u_i} + \frac{\partial}{\partial x_j}\left(-\rho \overline{u_{i'} u_{j'}}\right) \tag{2.7}$$

The additional term $-\rho \overline{u_{i'} u_{j'}}$ is known as the Reynolds stress tensor, which accounts for the apparent stresses introduced by turbulent fluctuations. These stresses are often several orders of magnitude larger than the molecular viscous stresses and are the principal reason why turbulent and laminar flow behaviors differ so significantly. Because Reynolds stresses cannot be computed directly from the RANS equations, additional turbulence models are required to close the system. These models attempt to relate the Reynolds stresses to mean flow quantities using empirical or semi-empirical formulations. The accuracy and applicability of such models vary depending on the flow configuration. Several widely used turbulence models will be discussed in the following section.

2.2.1 Turbulence Modeling Approaches

Turbulence models are generally categorized into three main classes based on the complexity and treatment of Reynolds stresses:

1. Algebraic eddy viscosity models,
2. Differential eddy viscosity models,
    - (a) One-equation models
    - (b) Two-equation or multi-equation models
3. Reynolds stress models (RSM)
    - Also known as second-moment closure models.

In both the algebraic and differential eddy viscosity models (categories 1 and 2), the Reynolds stresses are estimated using the Boussinesq hypothesis, which relates them to the mean strain rate via a turbulent (eddy) viscosity. These approaches assume isotropic turbulence and provide relatively simple formulations suitable for many engineering applications. In contrast, Reynolds Stress Models (category 3) involve solving separate transport equations for each component of the Reynolds stress tensor. This allows for the direct modeling of anisotropy in turbulence and offers improved predictive capabilities in complex flow situations, such as swirling flows, strong streamline curvature, and secondary motions.

2.2.2 *Boussinesq Hypothesis*

In 1877, Boussinesq proposed a model in which the apparent turbulent shear stress is expressed as the product of the mean velocity gradient and a turbulent viscosity coefficient, denoted by μ_t. This formulation is analogous to Newton's law of molecular viscosity, but with the distinction that μ_t is not constant and varies spatially.

$$\tau = -\rho\,\overline{u'v'} = \mu_t\,\frac{\partial \overline{u}}{\partial y} \tag{2.8}$$

The complete tensor form of the Boussinesq approximation for the Reynolds stress tensor is given by:

$$-\rho\,\overline{u_{i'}u_{j'}} = \mu_t\left(\frac{\partial \overline{u_i}}{\partial x_j} + \frac{\partial \overline{u_j}}{\partial x_i}\right) - \frac{2}{3}\rho k \delta_{ij} \tag{2.9}$$

where

- $u_{i'}$ and$u_{j'}$ are the fluctuating velocity components,
- μ_t is the eddy (turbulent) viscosity,
- $k = \frac{1}{2}\overline{u_i'u_i'}$ is the turbulent kinetic energy, and
- δ_{ij} is the Kronecker delta.

2.2.3 *Algebraic Turbulence Models*

In algebraic turbulence models, the turbulent velocity scale is estimated using the local velocity gradient, and a characteristic turbulence length scale (typically denoted l_{mix}) is used to define the size of energy-carrying eddies. The turbulent kinematic viscosity ν_t is typically computed from:

$$\nu_t = l_{\text{mix}}^2\,\frac{\partial u}{\partial y} \tag{2.10}$$

These models primarily aim to estimate the eddy viscosity based on flow gradients and a mixing length scale. The Reynolds stress tensor is then modeled using the Boussinesq hypothesis, which links turbulent stresses to the mean velocity gradient via μ_t. One of the most well-known algebraic models is the Prandtl Mixing Length Model, in which the turbulent viscosity is given by:

$$\nu_t = l_{\text{mix}}^2\left[\frac{\partial u_i}{\partial x_j}\left(\frac{\partial u_i}{\partial x_j} + \frac{\partial u_j}{\partial x_i}\right)\right]^{1/2} \tag{2.11}$$

The mixing length l_{mix} is typically determined empirically and varies depending on the flow conditions. For example:

- Near solid walls, Prandtl proposed:

$$l_{\text{mix}} \approx ky$$

  where y is the wall-normal distance and k is the von Kármán constant.

- For free-surface flows:

$$l_{\text{mix}} \approx C\sigma$$

  where C is an empirical constant and σ is the surface tension.

Shear-driven flows and pressure-driven internal flows are two examples of relatively simple flow scenarios that are best suited for algebraic turbulence models, which are sometimes thought of as the most fundamental class within turbulence modeling methodologies. These models are appealing for preliminary design studies, conceptual analysis, or low-cost simulations because they are computationally cheap and use empirical relations to approximate turbulent viscosity or mixing length (l_"mix"). Their ease of use makes it simple to incorporate them into numerical solvers, particularly when the geometry is simple and the flow is unaltered. But there are important drawbacks to this same simplicity. Complex geometries, flows with large curvature, or flows involving separation and reattachment are not well suited for algebraic models since it is very difficult and unreliable to define the local mixing length in these situations.

Furthermore, these models do not adequately capture the fundamental features of turbulence, especially in multi-dimensional flows where momentum diffusion, energy cascades, and cross-stream interactions are crucial. Their predictive value is limited for real-world applications involving secondary flows, severe pressure gradients, or unstable events because they are unable to represent anisotropic turbulence behavior or the redistribution of turbulent kinetic energy across directions. Their efficacy in engineering issues requiring high fidelity and precision is compromised by their incapacity to represent the physics of turbulent transport in a multidimensional framework. Consequently, algebraic models are best regarded as approximations, appropriate for initial assessments but inadequate for precisely resolving the entire complexity of turbulent transport in complex flow configurations, even though they offer a useful starting point for simulating turbulent flows.

In light of the above, details of characteristics of "Algebraic Models" include:

- Suitable for simple shear flows or pressure-driven internal flows.
- Computationally inexpensive and easy to implement.

- Not applicable for complex geometries or separated flows, where determining l_{mix} is difficult.
- Inadequate for representing turbulent transport properties such as energy and momentum diffusion in multi-dimensional flows.

2.2.4 One-Equation Differential Turbulence Models

In one-equation turbulence models, a single transport equation is solved to determine a characteristic turbulence quantity—typically the turbulent kinetic energy k. The second characteristic quantity, usually the turbulence length scale l_m, is then estimated algebraically. The turbulent viscosity ν_t is subsequently computed using the Boussinesq hypothesis.

A notable limitation of these models lies in the approximation of the length scale, which cannot be universally expressed by a single algebraic relation for all flow types. One of the earliest and most well-known models in this category is the Prandtl (1945) one-equation model. The governing equation for k in this model is:

$$\rho\frac{\partial k}{\partial t}+\rho\frac{\partial}{\partial x_i}\left(u_i k-\frac{\nu_t}{\sigma_k}\frac{\partial k}{\partial x_i}\right)=\rho(P_k+\Gamma_\delta-\varepsilon) \tag{12a}$$

with the following relations:

$$\nu_t = C_\mu\, k^{1/2}\, l_m \tag{12b}$$

$$\varepsilon=\frac{C_d\, k^{3/2}}{l_m} \tag{12c}$$

$$P_k=\nu_t\left(\frac{\partial u_i}{\partial x_j}\left(\frac{\partial u_i}{\partial x_j}+\frac{\partial u_j}{\partial x_i}\right)\right) \tag{12d}$$

$$\Gamma_\delta=-\frac{\nu_t\, g_i}{\rho\, \sigma_H}\frac{\partial P}{\partial x_i} \tag{12e}$$

where

- ν_t is the turbulent eddy viscosity,
- k is the turbulent kinetic energy,
- ε is the turbulent dissipation rate,
- l_m is the mixing length,
- g_i is the gravitational acceleration in direction x_i,
- σ_k and σ_H are the model constants, and
- $C_\mu = 0.5478$ and $C_d = 0.1643$ are the empirical constants.

In this formulation, Prandtl assumed the characteristic velocity scale to be proportional to $\sqrt{k}$, allowing for more accurate consideration of diffusion and convection effects compared to algebraic models. While this model performs well in boundary layer flows, it becomes insufficient for more complex flows due to its inability to capture length-scale variation dynamically.

As such, the application of one-equation models is generally limited to canonical wall-bounded flows, particularly boundary layers. In complex, three-dimensional, rotating, or swirling flows—such as those addressed in this study—one-equation models often fall short.

Therefore, advanced models that explicitly account for the transport and diffusion of both the velocity and length scales are required. These belong to the family of two-equation turbulence models, which will be introduced in the next section.

2.2.5 Two-Equation Differential Turbulence Models

As discussed in the previous sections, the modeling of complex turbulent flows requires accurate representation of both the turbulent velocity and length scales, including their transport and diffusion. This necessity has led to the development of two-equation turbulence models, in which transport equations are solved for both the turbulent kinetic energy and a second turbulence scale variable. Most two-equation models compute the eddy viscosity using the generalized relation:

$$\mu_t = \rho\, l\, k^{1/2} \tag{2.13}$$

In these models, transport equations are derived not only for the turbulent kinetic energy k, but also for a second variable z, which is a function of both the velocity and length scales. The general form of the second transport equation is given by:

$$\rho \frac{Dz}{Dt} = \frac{\partial}{\partial x_j}\left(\frac{\mu_t}{\sigma_z}\frac{\partial z}{\partial x_j}\right) + z\left\{C_1 \frac{\mu_t}{k}\left[\frac{\partial \overline{u}_i}{\partial x_j}\left(\frac{\partial \overline{u}_i}{\partial x_j} + \frac{\partial \overline{u}_j}{\partial x_i}\right)\right] - C_2 \frac{\rho^2 k}{\mu_t}\right\} + S_z \tag{2.14}$$

where

- z is the second turbulence quantity (e.g., dissipation rate ε, specific dissipation rate ω,, etc.),
- μ_t is the eddy viscosity,
- σ_z is the turbulent Prandtl number for z,
- $C_1 and C_2$ are the model constants,
- S_z is the secondary source term, and
- $\overline{u}_i$ is the mean velocity components.

The general procedure for constructing such models involves the following steps:

1. Definition of the second variable $Z = k^m l^n$, where m and n are the constants.
2. Derivation of a transport equation for Z from the Navier–Stokes equations.
3. Closure of unknown terms using empirical or semi-empirical relations.

Once the governing equations for k and Z are solved, the eddy viscosity μ_t can be evaluated using the relation above. It is important to note that, unlike one-equation models where the length scale l is estimated directly, here Z is a composite variable from which l is indirectly determined. Depending on how Z is formulated, various two-equation turbulence models arise. Among these, the k–ε model has gained widespread application across many engineering disciplines due to its robustness and simplicity. In this study, a modified version known as the RNG k–ε model—a statistically derived variant that improves accuracy for swirling and separated flows—has been adopted for the analysis of complex turbulent regimes.

2.2.6 The k–ε RNG Turbulence Model

In this model, the auxiliary variable Z is defined as $\frac{k^{3/2}}{l}$ and is denoted by ε, which represents the dissipation rate of turbulent kinetic energy. The dissipation rate is given by:

$$\varepsilon = \nu \overline{\frac{\partial u_{i'}}{\partial x_j} \frac{\partial u_{i'}}{\partial x_j}} \tag{2.15}$$

Based on the Navier–Stokes equations, the tensor form of the steady-state transport equations for the turbulent kinetic energy k and the dissipation rate ε can be expressed as follows [5]:

Turbulent Kinetic Energy Equation:

$$\frac{\partial(\rho \bar{u}_i k)}{\partial x_i} = \frac{\partial}{\partial x_j}\left(\frac{\mu_{\text{eff}}}{\sigma_k}\frac{\partial k}{\partial x_j}\right) + \mu_t S^2 - \rho\varepsilon \tag{2.16}$$

Dissipation Rate Equation:

$$\frac{\partial \rho \bar{u}_i \varepsilon}{\partial x_i} = \frac{\partial}{\partial x_i}\left(\frac{\mu_{\text{eff}}}{\sigma_\varepsilon}\frac{\partial \varepsilon}{\partial x_i}\right) + C_{1\varepsilon}\frac{\mu_t S^2 \varepsilon}{k} - C_{2\varepsilon}\frac{\rho\varepsilon^2}{k} - R \tag{2.17}$$

where

- μ_t is the turbulent (eddy) viscosity,
- $\mu_{\text{eff}} = \mu + \mu_t$ is the effective viscosity,
- $S = \sqrt{2S_{ij}S_{ij}}$, with $S_{ij} = \frac{1}{2}\left(\frac{\partial \bar{u}_i}{\partial x_j} + \frac{\partial \bar{u}_j}{\partial x_i}\right)$ is the mean strain rate tensor,

- $\eta = \frac{Sk}{\varepsilon}$ is the dimensionless strain parameter, and
- $R = \frac{C_\mu \rho \eta^3 (1-\eta/\eta_0)}{1+\beta\eta^3} \cdot \frac{\varepsilon^2}{k}$ is the additional term introduced in RNG formulation.

The model constants are:

- $C_\mu = 0.0845$,
- $C_{1\varepsilon} = 1.42$, $C_{2\varepsilon} = 1.68$,
- $\sigma_k = \sigma_\varepsilon = 0.718$,
- $\eta_0 = 4.38$, $\beta = 0.011$–0.015.

Equation (2.16) can be re-expressed to include the modified dissipation coefficient:

$$\frac{\partial(\rho \overline{u}_i \varepsilon)}{\partial x_i} = \frac{\partial}{\partial x_i}\left(\frac{\mu_{\text{eff}}}{\sigma_\varepsilon}\frac{\partial \varepsilon}{\partial x_j}\right) + C_{1\varepsilon}\frac{\mu_t S^2 \varepsilon}{k} - C^*_{2\varepsilon}\frac{\rho \varepsilon^2}{k} \tag{2.18}$$

where the modified coefficient $C^*_{2\varepsilon}$ is given by:

$$C^*_{2\varepsilon} = C_{2\varepsilon} + \frac{C_\mu \eta^3 (1-\eta/\eta_0)}{1+\beta\eta^3} \tag{2.19}$$

The RNG (Renormalization Group) variant of the k–ε model derives its production and dissipation coefficients from analytical procedures rather than empirical calibration. The inclusion of the term R in the dissipation equation enhances the model's ability to resolve inhomogeneous turbulence structures, particularly in regions of high strain rate.

In low-strain regions where $\eta < \eta_0$, the correction term R is positive, causing $C^*_{2\varepsilon}$ to exceed the standard $C_{2\varepsilon}$ (typically 1.92 in the standard k–ε model). Therefore, in such flows, the RNG and standard models produce similar results. However, in strongly strained flow regions, R becomes negative, reducing $C^*_{2\varepsilon}$ and thereby enhancing the accuracy of the RNG formulation over the standard model [6].

2.2.7 Wall Function Implementation for the k–ε RNG Model

In this study, the *k–ε RNG* model employs the wall function approach to satisfy boundary conditions at solid walls. Prior to discussing the wall function methodology, it is essential to note that the turbulent flow field near a wall is typically divided into the following three regions:

- Laminar sublayer: The region where molecular viscosity dominates over turbulence effects.
- Buffer zone: A transitional layer where both molecular and turbulent effects are of comparable magnitude.
- Fully turbulent core: A region where turbulent transport significantly exceeds molecular diffusion.

To determine which region a given near-wall point belongs to, the dimensionless wall distance, denoted by y^+, is computed using the following expression:

$$y^+ = \frac{(C_D C_\mu)^{0.25} k_p^{0.5} y_p}{\nu} \tag{2.20}$$

where

- k_p is the turbulent kinetic energy at the first grid point near the wall,
- y_p is the normal distance from the wall to this point,
- ν is the kinematic viscosity,
- C_D and C_μ are the turbulence model constants.

The flow behavior at the near-wall point is categorized as follows:

- For $y^+ < 5$, the point lies within the laminar sublayer, where the velocity profile is linear:

$$\frac{\overline{u}}{u_\tau} = y^+ \tag{2.21}$$

with u_τ being the friction velocity, defined by:

$$u_\tau = \sqrt{\frac{\tau_w}{\rho}} \tag{2.22}$$

where τ_w is the wall shear stress, and ρ is the fluid density.

- For $5 < y^+ < 30$, the point resides in the buffer zone, where the velocity profile is given by the empirical expression:

$$\frac{\overline{u}}{u_\tau} = 4.2 - 5.7\ln(y^+) + 5.1(\ln(y^+))^2 - 0.7(\ln(y^+))^3 \tag{2.23}$$

- For $30 < y^+ < 1000$, the point falls within the fully turbulent core, where the logarithmic law of the wall applies:

$$\frac{\overline{u}}{u_\tau} = \kappa \ln(E y^+) \tag{2.24}$$

Here,

- $K = 0.418$ is the von Kármán constant, and
- $E = 9.758$ is a constant that can be modified based on wall roughness.

The no-slip condition at the wall is implicitly satisfied through the wall function approach, which ensures accurate evaluation of turbulent shear stress at the first grid point adjacent to the wall. The process involves:

1. Calculating y^+ using Eq. (2.20) to identify the appropriate near-wall region.
2. Applying the corresponding velocity profile (Eqs. 2.21–2.23) to determine the near-wall boundary condition for the velocity component.
3. Imposing boundary conditions for the k and ε equations based on the assumption that the production and dissipation of turbulent kinetic energy are balanced at the near-wall point.

The advantage of using wall functions in the *k–ε RNG* model lies in the ability to employ coarser grids near solid boundaries, thus significantly reducing computational cost and simulation time. Furthermore, the model can easily account for surface roughness effects through the parameter E, which varies as a function of roughness characteristics.

2.3 Heat Transfer Equations

During the daytime, the temperature of building roofs and walls exposed to solar radiation increases significantly. As a result of convective heat transfer caused by wind and thermal exchange with the sky, which generally has a lower temperature than the surrounding atmosphere, the temperature of these surfaces undergoes partial reduction. Simultaneously, heat is conducted from the exterior surfaces through the building envelope into the interior spaces. Under high solar radiation conditions, this conductive transfer can significantly elevate the cooling load within the building [3, 4]. To evaluate the spatial and temporal distribution of temperature and heat transfer, it is essential to solve the governing energy equations.

In this chapter, the heat transfer mechanisms considered include forced convection, thermal conduction, solar radiation, and longwave radiative exchange with the sky. The governing equations corresponding to each of these processes are provided in the subsequent subsections. These formulations serve as the foundation for evaluating thermal behavior in urban environments and built structures. Particular attention is given to the interactions between surface properties and environmental conditions, which critically influence heat exchange dynamics.

2.3.1 Energy Equation for Turbulent Flow

The general form of the unsteady energy conservation equation in a compressible or incompressible flow field is expressed as:

$$\rho C_p \frac{\partial T}{\partial t} + \frac{\partial}{\partial x_i}(\rho C_p u_i T) = \frac{\partial}{\partial x_i}\left(k \frac{\partial T}{\partial x_i}\right) + \frac{Dp}{Dt} + \mu \Phi \tag{2.25}$$

where

- ρ is the fluid density [kg/m^3],
- C_p is the specific heat capacity at constant pressure [J/(kg·K)],
- T is the temperature [K],
- u_i is the velocity component in the x_i-direction [m/s],
- k is the thermal conductivity [W/(m·K)],
- p is the pressure [Pa],
- μ is the dynamic viscosity [Pa·s], and
- Φ is the viscous dissipation function [W/m^3].

Assuming incompressible flow, constant thermophysical properties, and negligible pressure work and viscous dissipation, the equation simplifies significantly. When turbulence effects are introduced using Reynolds averaging and the Boussinesq hypothesis, the energy equation transforms into the following Reynolds-averaged form:

$$\frac{\partial T}{\partial t} + \frac{\partial (u_i T)}{\partial x_i} = \frac{\partial}{\partial x_i}\left[\left(\frac{\lambda}{\rho C_p} + \frac{\varepsilon_m}{\mathrm{Pr}_t}\right)\frac{\partial T}{\partial x_i}\right] \tag{2.26}$$

Here

- λ is the molecular thermal conductivity [W/(m·K)],
- ε_m is the eddy viscosity (turbulent dynamic viscosity) [kg/(m·s)], and
- Pr_t is the turbulent Prandtl number.

This equation accounts for both molecular and turbulent contributions to thermal diffusion and is essential for modeling energy transport in turbulent atmospheric flows or indoor environments subjected to wind-induced ventilation.

2.3.2 *Heat Conduction Equation in the Roof Structure*

Assuming spatially and temporally constant material properties, the transient one-dimensional heat conduction equation through the roof layer is given by:

$$\rho C_{p_r} \frac{\partial T_r}{\partial t} = \frac{\partial}{\partial x_i}\left(\lambda_r \frac{\partial T_r}{\partial x_i}\right) \tag{2.27}$$

where

- T_r is the temperature distribution within the roof material [K],
- C_{p_r} is the specific heat capacity of the roof material [J/(kg·K)], and
- λ_r is the thermal conductivity of the roof material [W/(m·K)].

This equation forms the basis for assessing conductive heat transfer through building envelopes under dynamic boundary conditions such as solar loading and convective cooling.

2.3.3 Solar Radiation

Solar radiation consists of three main components:

1. Direct Beam Radiation: This component represents the portion of solar radiation that reaches a surface directly from the sun, without atmospheric scattering. It is only present when the surface has a direct line of sight to the sun.
2. Diffuse Radiation: This component arises from the scattering of sunlight by atmospheric particles and gas molecules. Its magnitude increases significantly in the presence of cloud cover or air pollution. In fully overcast conditions, the diffuse component dominates while the direct component becomes minimal.
3. Reflected Radiation: This is the portion of solar radiation reflected from surrounding surfaces such as the ground or adjacent buildings. The reflectivity depends on the type of ground cover and material properties of the reflecting surfaces.

Several empirical and semi-empirical models are available to estimate the direct and diffuse components of solar radiation, including the Daneshyar model [5], the ASHRAE model, and the Hottel model [6].

The Daneshyar model can be selected due to its proven suitability for the climatic conditions prevalent in Iran. Before applying this model, it is essential to define a set of solar geometric parameters that determine the position of the sun in the sky and the orientation of the surface receiving radiation. These parameters, illustrated in Fig. 2.1, are described below:

- Φ (Latitude): The geographic latitude of the location under study.
- δ (Declination Angle): The angular position of the sun at solar noon relative to the equatorial plane.
- β (Surface Slope): The angle between the inclined surface and the horizontal plane.
- γ (Surface Azimuth Angle): The angle between the projection of the surface normal vector on the horizontal plane and the true south direction.
- ω (Hour Angle): The angular displacement of the sun from the local solar noon, expressed in degrees (15° per hour).
- θ (Incidence Angle): The angle between the incoming solar rays and the normal vector to the surface.
- θ_z (Zenith Angle): The angle between the sun's rays and the vertical direction.

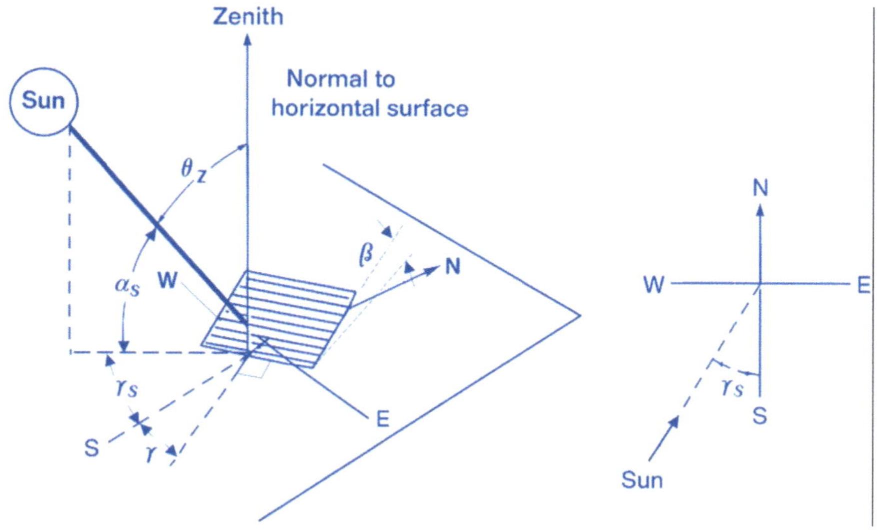

Fig. 2.1 Solar Angles Used in the Calculation of Incident Solar Radiation

- α_s (Solar Altitude Angle): The elevation angle of the sun above the horizontal plane.
- γ_s (Solar Azimuth Angle): The angle between the projection of the sun's rays onto the horizontal plane and the south direction.

The angle of incidence of direct solar radiation on a given surface is determined by the following expression:

$$\cos\theta = \cos\theta_z \cos\beta + \sin\theta_z \cos(\gamma_s - \gamma) \tag{2.28}$$

Each of the angles involved in Eq. (2.27) can be calculated from the relations below:

$$\cos\theta_z = \cos\phi \cos\delta \cos\omega + \sin\phi \sin\delta \tag{2.29}$$

To determine the solar azimuth angle γ_s, it is necessary to first determine the hemisphere in which the sun is located, based on the relationship between the hour angle ω and the sunset hour angle ω_{ew}:

$$\gamma_s = C_1 C_2 \gamma_{s'} + C_3 \left(\frac{1 - C_1 C_2}{2} \right) \times 180^\circ \tag{2.30}$$

$$\sin\gamma_{s'} = \frac{\sin\omega \cos\delta}{\sin\theta_z} \tag{2.31}$$

The constants C_1, C_2, and C_3 are defined as follows:

$$C_1 = \begin{cases} 1 & \text{if } |\omega| \le \omega_{ew} \\ -1 & \text{if } |\omega| > \omega_{ew} \end{cases}$$

$$C_2 = \begin{cases} 1 & \text{if } (\phi - \delta) \ge 0 \\ -1 & \text{if } (\phi - \delta) < 0 \end{cases}$$

$$C_3 = \begin{cases} 1 & \text{if } \omega \ge 0 \\ -1 & \text{if } \omega < 0 \end{cases}$$

The sunset hour angle ω_{ew}, which indicates the angle of sunset relative to solar noon, is computed as:

$$\cos\omega_{ew} = \frac{\tan\delta}{\tan\phi} \tag{2.32}$$

The solar declination angle δ, which represents the angular position of the sun at solar noon with respect to the celestial equator, is given by:

$$\delta = 23.45^\circ \sin\left(\frac{360^\circ}{365}(284 + n)\right) \tag{2.33}$$

where n is the day number of the year (e.g., September 1 corresponds to $n = 244$).

The sunset (or sunrise) hour angle ω_s, i.e., the angular displacement of the sun from solar noon at sunset or sunrise, is determined from:

$$\cos\omega_s = -\tan\phi \tan\delta \tag{2.34}$$

Accordingly, the total day length in hours N is calculated as:

$$N = \frac{2}{15}\cos^{-1}(\tan\phi \tan\delta) \tag{2.35}$$

Having now defined the essential geometric and astronomical parameters required for solar radiation estimation, the Daneshyar model—which is suitable for the climatic conditions of Iran—can be introduced and applied in the following section.

2.3.4 Daneshyar Radiation Model

The Daneshyar model [5] provides an empirical framework for estimating the components of solar radiation, particularly under the climatic conditions of Iran. In this model, the direct solar radiation component is calculated using a modified average extraterrestrial radiation G_c, which varies depending on geographic location. The direct beam irradiance incident on a tilted surface is given by:

$$G_{\text{direct}} = \left(1 - C_f\right)G_c\left[1 - \exp(-4.2972\alpha_s)\right]\cos\theta \quad (2.36)$$

where

- G_{direct} is the direct solar radiation component (W/m^2),
- C_f is the cloudiness factor (dimensionless), varies from 0 (clear sky) to 1 (overcast),
- G_c is the average extraterrestrial solar irradiance; for Iran, Daneshyar recommends $G_c = 950.61\ W/m^2$, consistent with values used for Australia,
- α_s is the solar altitude angle (radians), and
- θ is the solar incidence angle on the surface (degrees).

The cloud factor C_f accounts for atmospheric effects such as humidity, aerosol concentration, and air pollution and has been empirically estimated for different Iranian cities using ten-year meteorological data.

The diffuse solar radiation component is calculated using a regression-based expression:

$$G_{\text{diffuse}} = \left(A_1 + A_2\alpha_s + A_3C_f\right)F_{\text{ws}} \quad (2.37)$$

where

- G_{diffuse} is the diffuse radiation component (W/m^2), and
- F_{ws} is the sky view factor, representing the fraction of sky visible from the surface.

The empirical coefficients adapted for Iranian climatic conditions are as follows:

$$A_1 = 1.5352A_2 = 120.6093A_3 = 121.301$$

The sky view factor F_{ws} is determined from the orientation angle of the surface (β) using the following equation:

$$F_{\text{ws}} = \frac{1 + \cos\beta}{2} \quad (2.38)$$

where

- β: Inclination angle of the surface from the horizontal (°).

- o For a horizontal surface: $\beta = 0° \Rightarrow F_{\text{ws}} = 1$
- o For a vertical surface: $\beta = 90° \Rightarrow F_{\text{ws}} = 0.5$

The reflected radiation component G_R is modeled as the ground-reflected portion of the total incoming solar radiation, expressed as:

$$G_R = G_{\text{th}} \cdot \rho_g \cdot F_{\text{wg}} \tag{2.39}$$

where

- G_R is the reflected radiation (W/m^2),
- G_{th} is the total incident solar radiation on the ground (W/m^2),
- ρ_g is the ground reflectivity (dimensionless), and
- F_{wg} is the view factor between the vertical surface (e.g., wall) and the ground.

For a planar vertical wall, the ground view factor is given by:

$$F_{\text{wg}} = \frac{1 - \cos\beta}{2} \tag{2.40}$$

It follows from Eqs. (2.2–2.37) and (2–39) that for a vertical wall ($\beta = 90°$), the sky and ground view factors are equal, each yielding:

$$F_{\text{ws}} = F_{\text{wg}} = \frac{1}{2}$$

This model provides a physically consistent approach to estimating the three solar radiation components (direct, diffuse, and reflected) and is especially suited for building energy simulations and outdoor thermal comfort analyses in arid and semi-arid climates.

2.3.5 *Radiative Heat Exchange with the Sky*

In arid and desert regions, where the sky is predominantly clear, the radiative heat exchange with the sky—which typically has a temperature lower than that of the ambient air—plays a significant role in reducing the temperature of terrestrial surfaces. This radiative cooling effect becomes particularly important for energy balance modeling in urban microclimates and building thermal simulations.

The net radiative heat flux between a horizontal surface and the sky is given by the Stefan–Boltzmann law:

$$q = \sigma\varepsilon\left(T_{\text{sur}}^4 - T_{\text{sky}}^4\right) \tag{2.41}$$

where

- q is the net longwave radiation heat flux (W/m^2),
- σ is the Stefan–Boltzmann constant; $\sigma = 5.67 \times 10^{-8} \frac{\mathrm{W}}{\mathrm{m^2K^4}}$,
- ε is the emissivity of the surface (dimensionless); assumed to be 0.8,
- T_{sur} is the surface temperature (K), and
- T_{sky} is the effective sky temperature (K).

The sky temperature, T_{sky}, can be empirically estimated as a function of the ambient air temperature T_{amb}, using the relation proposed in [11]:

$$T_{\mathrm{sky}} = 1.57 \times T_{\mathrm{amb}}^{1/3} \tag{2.42}$$

where

- T_{amb} is the ambient air temperature (K).

The diurnal variation of ambient temperature can be modeled using a cosine-based periodic function, commonly employed in thermal modeling of outdoor conditions:

$$T_{\mathrm{amb}} = \frac{T_{\mathrm{max}} + T_{\mathrm{min}}}{2} + \frac{T_{\mathrm{max}} - T_{\mathrm{min}}}{2}\cos\left(\frac{\pi(t-15)}{12}\right) \tag{2.43}$$

where

- T_{max} is the maximum daily air temperature (K),
- T_{min} is the minimum daily air temperature (K), and
- t is the time of day in hours (h), with $t = 15$ corresponding to 3:00 PM (typically the warmest hour)

This formulation captures the periodic nature of ambient temperature over a 24-h cycle and serves as input for estimating sky temperature and subsequent radiative heat exchange.

In the subsequent section, we shall derive the non-dimensional forms of the governing equations, which facilitate numerical implementation and parameter analysis in CFD simulations.

2.4 Non-Dimensional Form of the Governing Equations

In fluid flow analysis, adopting the non-dimensional form of the governing equations offers significant advantages. It enables the generalization of a wide range of flow configurations under a unified framework and eliminates the need for solving similar cases repeatedly. In this section, we first define the relevant non-dimensional parameters, followed by the transformation of the governing equations into their dimensionless form.

2.4.1 Non-Dimensional Parameters

$$x^* = \frac{x}{L} \tag{2.44}$$

$$y^* = \frac{y}{L} \tag{2.45}$$

$$t^* = \frac{t u_{\mathrm{in}}}{L} \tag{2.46}$$

$$u^* = \frac{u}{u_{\mathrm{in}}} \tag{2.47}$$

$$v^* = \frac{v}{u_{\mathrm{in}}} \tag{2.48}$$

$$p^* = \frac{p}{\rho_{\mathrm{in}} u_{\mathrm{in}}^2} \tag{2.49}$$

$$\theta = \frac{T}{T_{\mathrm{in}}} \tag{2.50}$$

$$\mu_t^* = \frac{\mu_t}{\mu} \tag{2.51}$$

$$k^* = \frac{k}{u_{\mathrm{in}}^2} \tag{2.52}$$

$$\varepsilon^* = \frac{\varepsilon}{u_{\mathrm{in}}^3 / L} \tag{2.53}$$

$$\rho^* = \frac{\rho}{\rho_{\mathrm{in}}} \tag{2.54}$$

$$c^* = \frac{c}{c_{\mathrm{in}}} \tag{2.55}$$

$$\lambda^* = \frac{\lambda}{\lambda_{\mathrm{ib}}} \tag{2.56}$$

2.4.2 Dimensionless Governing Equations

Continuity Equation:

$$\frac{\partial \overline{u}_i^*}{\partial x^* i} = 0 \tag{2.57}$$

Momentum Equation:

$$\frac{\partial \overline{u}_i^*}{\partial t^*} + \overline{u}_j^* \frac{\partial \overline{u}_i^*}{\partial x_j^*} = -\frac{\partial \overline{p}^*}{\partial x_i^*} + \frac{\partial}{\partial x_j^*}\left[\frac{1+\mu_t^*}{Re}\frac{\partial \overline{u}_i^*}{\partial x_j^*}\right] + \frac{\partial}{\partial x_j^*}\left[\frac{\mu_t^*}{Re}\frac{\partial \overline{u}_j^*}{\partial x_i^*}\right] \tag{2.58}$$

Turbulent Kinetic Energy Equation (RNG k–ε Model):

$$\frac{\partial k^*}{\partial t^*} + \frac{\partial\left(u_i^* k^*\right)}{\partial x_i^*} = \frac{\partial}{\partial x_i^*}\left[\frac{1+\mu_t^*/\sigma_k}{Re}\frac{\partial k^*}{\partial x_i^*}\right] + \frac{\mu_t^* G^*}{Re} - C_D \varepsilon^* \tag{2.59}$$

Turbulent Dissipation Rate Equation (RNG k–ε Model):

$$\frac{\partial \varepsilon^*}{\partial t^*} + \frac{\partial\left(u_i^* \varepsilon^*\right)}{\partial x_i^*} = \frac{\partial}{\partial x_i^*}\left[\frac{1+\mu_t^*/\sigma_\varepsilon}{Re}\frac{\partial \varepsilon^*}{\partial x_i^*}\right] + \frac{C_{1\varepsilon}}{Re}\mu_t^* G^* \frac{\varepsilon^*}{k^*} - C_{2\varepsilon}^* \frac{\varepsilon^{*2}}{k^*} \tag{2.60}$$

Production Term of Turbulent Kinetic Energy:

$$G^* = 2\left(\frac{\partial \overline{u}^*}{\partial x^*}\right)^2 + 2\left(\frac{\partial \overline{v}^*}{\partial y^*}\right)^2 + \left(\frac{\partial \overline{u}^*}{\partial y^*} + \frac{\partial \overline{v}^*}{\partial x^*}\right)^2 \tag{2.61}$$

Modified Dissipation Coefficient:

$$C_{2\varepsilon}^* = C_{2\varepsilon} + C_\mu \frac{\eta^3(1-\eta/\eta_0)}{1+\beta\eta^3} \tag{2.62}$$

Turbulent Viscosity:

$$\mu_t^* = \mathrm{Re} \cdot C_\mu \cdot \frac{K^{*2}}{\varepsilon^*} \tag{2.63}$$

2.4.3 Dimensionless Heat Transfer Equations

Convective Heat Transfer Equation:

$$\frac{\partial\theta}{\partial t^*}+\frac{\partial\left(\overline{u}_i^*\theta\right)}{\partial x_i^*}=\frac{\partial}{\partial x_i^*}\left[\left(\frac{1}{RePr}+\frac{\mu_t^*}{RePr_t}\right)\frac{\partial\theta}{\partial x_i^*}\right] \tag{2.64}$$

Conduction Heat Transfer in Roof Structure:

$$\rho_r^* c_r^* \frac{\partial\theta_r}{\partial t^*}=\frac{1}{Re\cdot Pr}\frac{\partial}{\partial x_i^*}\left(\lambda_r^*\frac{\partial\theta_r}{\partial x_i^*}\right) \tag{2.65}$$

2.5 Coordinate System Transformation

In curvilinear coordinate systems, two types of basis vectors are defined. One type is tangent to the coordinate lines and defines the directions of the coordinate axes. These are referred to as covariant basis vectors, denoted by $e_{(i)}$. The other type, orthogonal to the coordinate surfaces, is known as contravariant basis vectors, denoted by $e^{(i)}$. These vectors are illustrated in Fig. 2.2.

The covariant basis vectors are defined as:

$$\boldsymbol{e}_{(i)}=\frac{\partial\boldsymbol{x}}{\partial\xi^i} \tag{2.66}$$

while the contravariant basis vectors are expressed as:

$$e^{(i)}=\frac{\partial\xi^i}{\partial x} \tag{2.67}$$

Here, $\boldsymbol{e}_i$ represents a unit tangent vector (covariant), and $\boldsymbol{e}^i$ corresponds to a unit normal vector (contravariant). The chain rule that relates two curvilinear coordinate systems in 2D, namely $\xi=\xi(xyt)$ and $\eta=\eta(xy\tau)$, is given by [12]:

$$\frac{\partial}{\partial t}=\frac{\partial\tau}{\partial t}\frac{\partial}{\partial\tau}+\frac{\partial\xi}{\partial t}\frac{\partial}{\partial\xi}+\frac{\partial\eta}{\partial t}\frac{\partial}{\partial\eta}=\tau_t\frac{\partial}{\partial\tau}+\xi_t\frac{\partial}{\partial\xi}+\eta_t\frac{\partial}{\partial\eta} \tag{2.68}$$

$$\frac{\partial}{\partial x}=\frac{\partial\tau}{\partial x}\frac{\partial}{\partial\tau}+\frac{\partial\xi}{\partial x}\frac{\partial}{\partial\xi}+\frac{\partial\eta}{\partial x}\frac{\partial}{\partial\eta}=\tau_x\frac{\partial}{\partial\tau}+\xi_x\frac{\partial}{\partial\xi}+\eta_x\frac{\partial}{\partial\eta} \tag{2.69}$$

$$\frac{\partial}{\partial y}=\frac{\partial\tau}{\partial y}\frac{\partial}{\partial\tau}+\frac{\partial\xi}{\partial y}\frac{\partial}{\partial\xi}+\frac{\partial\eta}{\partial y}\frac{\partial}{\partial\eta}=\tau_y\frac{\partial}{\partial\tau}+\xi_y\frac{\partial}{\partial\xi}+\eta_y\frac{\partial}{\partial\eta} \tag{2.70}$$

Fig. 2.2 Covariant and Contravariant Basis Vectors

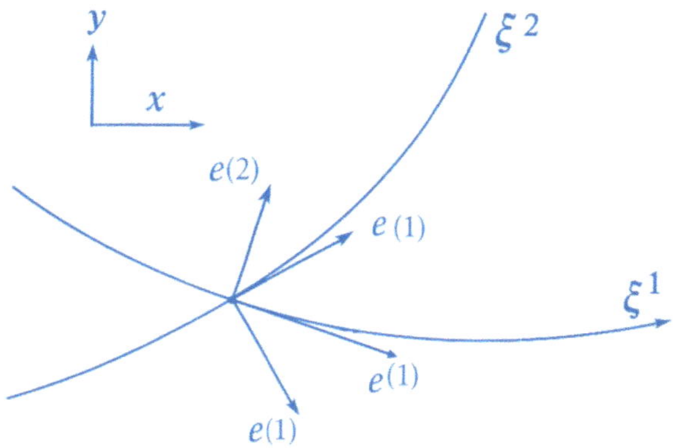

The surface area formed by two of the covariant basis vectors $\boldsymbol{e}_{(1)}\boldsymbol{e}_{(2)}$ and $\boldsymbol{e}_{(3)}$ is:

$$A^{(k)} = e_{(l)} \times e_{(m)} \tag{2.71}$$

where kl and m are the cyclic indices. Introducing the coordinate parameters ξ and η in the directions of $\boldsymbol{e}_{(1)}$ and $\boldsymbol{e}_{(2)}$, the Jacobian determinant J is defined as:

$$J = \begin{bmatrix} t_\tau & t_\xi & t_\eta \\ x_\tau & x_\xi & x_\eta \\ y_\tau & y_\xi & y_\eta \end{bmatrix} = \begin{bmatrix} 1 & 0 & 0 \\ x_\tau & x_\xi & x_\eta \\ y_\tau & y_\xi & y_\eta \end{bmatrix} = x_\xi y_\eta - x_\eta y_\xi \tag{2.72}$$

A control volume of dimensions $\delta\xi\delta\eta$ and $\delta\zeta$ in computational space is bounded by vectors along $\delta\xi\ e^{(1)}, \delta\eta\ e^{(2)}, and, \delta\zeta\ e^{(3)}$. Hence, the area and volume elements are given as:

$$\delta A^{(k)} = A^{(k)}\delta\xi^l\delta\xi^m \tag{2.73}$$

$$\delta V = J\ \delta\xi\ \delta\eta \tag{2.74}$$

The general form of the conservation equations for mass, momentum, energy, and turbulence is expressed as:

$$\frac{\partial\phi}{\partial t} + \frac{\partial(u\phi)}{\partial x} + \frac{\partial(v\phi)}{\partial y} = \frac{\partial}{\partial x}\left(\Gamma\frac{\partial\phi}{\partial x}\right) + \frac{\partial}{\partial y}\left(\Gamma\frac{\partial\phi}{\partial y}\right) + S \tag{2.75}$$

Here, ϕ is a generic scalar variable, Γ is the diffusion coefficient, and S represents the source term.

The metric coefficients in curvilinear coordinates are defined as:

$$\xi_t = \frac{-y_\eta x_\tau + x_\eta y_\tau}{J} \tag{2.76}$$

$$\xi_x = \frac{y_\eta}{J} \tag{2.77}$$

$$\xi_y = \frac{-x_\eta}{J} \tag{2.78}$$

$$\eta_t = \frac{y_\xi x_\tau - x_\xi y_\tau}{J} \tag{2.79}$$

$$\eta_x = \frac{-y_\xi}{J} \tag{2.80}$$

$$\eta_y = \frac{x_\xi}{J} \tag{2.81}$$

By substituting the chain rule expressions in Eqs. (2.55)–(2.57) into the general transport equation (Eq. 2.62), the transformed equation in curvilinear coordinates takes the following form:

$$\frac{\partial(J\phi)}{\partial\tau} - \frac{\partial(X\phi)}{\partial\xi} - \frac{\partial(Y\phi)}{\partial\eta} + \frac{\partial(U\phi)}{\partial\xi} + \frac{\partial(V\phi)}{\partial\eta}$$
$$= \frac{\partial}{\partial\xi}\left(\frac{\Gamma}{J}\left[c_1\frac{\partial\phi}{\partial\xi} - c_2\frac{\partial\phi}{\partial\eta}\right]\right) + \frac{\partial}{\partial\eta}\left(\frac{\Gamma}{J}\left[-c_2\frac{\partial\phi}{\partial\xi} + c_3\frac{\partial\phi}{\partial\eta}\right]\right) + JS \tag{2.82}$$

where the transformed convective flux coefficients are defined as:

$$X = y_\eta x_\tau - x_\eta y_\tau \tag{2.83}$$

$$Y = -y_\xi x_\tau + x_\xi y_\tau \tag{2.84}$$

$$U = y_\eta u - x_\eta v \tag{2.85}$$

$$V = -y_\xi u + x_\xi v \tag{2.86}$$

$$c_1 = x_\eta^2 + y_\eta^2 \tag{2.87}$$

$$c_2 = x_\xi x_\eta + y_\xi y_\eta \tag{2.88}$$

$$c_3 = x_\xi^2 + y_\xi^2 \tag{2.89}$$

By specifying the appropriate forms of ϕ, Γ, and S, Eq. (2.64) yields the governing equations for continuity, momentum, and energy in generalized curvilinear coordinates.

Continuity Equation

Setting $\Gamma = 0$, $\phi = 1$, and $S = 0$, the transformed continuity equation becomes:

$$\frac{\partial U}{\partial \xi} + \frac{\partial V}{\partial \eta} = 0 \tag{2.90}$$

Momentum Equation in the x-Direction

For $\phi = u$, $\Gamma = \frac{1+\mu_t}{Re}$, and assuming pressure gradients are present, the momentum equation in the x-direction is expressed as:

$$\begin{aligned}
\frac{\partial (Ju)}{\partial \tau} + \frac{\partial Uu}{\partial \xi} + \frac{\partial Vu}{\partial \eta} &= -y_\eta \frac{\partial p}{\partial \xi} + y_\xi \frac{\partial p}{\partial \eta} + \frac{\partial}{\partial \xi}\left[\frac{1+\mu_t}{Re}\left(\frac{c_1}{J}\frac{\partial u}{\partial \eta} - \frac{c_2}{J}\frac{\partial u}{\partial \eta}\right)\right] \\
&+ \frac{\partial}{\partial \eta}\left[\frac{1+\mu_t}{Re}\left(\frac{c_3}{J}\frac{\partial u}{\partial \eta} - \frac{c_2}{J}\frac{\partial u}{\partial \xi}\right)\right] \\
&+ \frac{\partial}{\partial \xi}\left[\frac{\mu_t}{Re}\left[\frac{y_\eta^2}{J}\frac{\partial u}{\partial \xi} - \frac{y_\eta y_\xi}{J}\frac{\partial u}{\partial \eta} - \frac{y_\eta x_\eta}{J}\frac{\partial v}{\partial \xi} + \frac{x_\eta y_\xi}{J}\frac{\partial v}{\partial \eta}\right)\right] \\
&+ \frac{\partial}{\partial \eta}\left[\frac{\mu_t}{Re}\left[-\frac{y_\xi y_\eta}{J}\frac{\partial u}{\partial \xi} + \frac{y_\xi^2}{J}\frac{\partial u}{\partial \eta} + \frac{x_\xi y_\eta}{J}\frac{\partial v}{\partial \xi} + \frac{x_\xi y_\xi}{J}\frac{\partial v}{\partial \eta}\right)\right]
\end{aligned} \tag{2.91}$$

Momentum Equation in the y-Direction.

For $\phi = v$, $\Gamma = \frac{1+\mu_t}{Re}$, the momentum equation in the y-direction becomes:

$$\begin{aligned}
\frac{\partial (Jv)}{\partial \tau} + \frac{\partial Uv}{\partial \xi} + \frac{\partial Vv}{\partial \eta} &= -x_\eta \frac{\partial p}{\partial \xi} - x_\xi \frac{\partial p}{\partial \eta} + \frac{\partial}{\partial \xi}\left[\frac{1+\mu_t}{Re}\left(\frac{c_1}{J}\frac{\partial v}{\partial \xi} - \frac{c_2}{J}\frac{\partial v}{\partial \eta}\right)\right] \\
&+ \frac{\partial}{\partial \eta}\left[\frac{1+\mu_t}{Re}\left(\frac{c_3}{J}\frac{\partial v}{\partial \eta} - \frac{c_2}{J}\frac{\partial v}{\partial \xi}\right)\right] \\
&+ \frac{\partial}{\partial \xi}\left[\frac{\mu_t}{Re}\left[-\frac{y_\eta x_\eta}{J}\frac{\partial u}{\partial \xi} - \frac{y_\eta x_\xi}{J}\frac{\partial u}{\partial \eta} - \frac{x_\eta^2}{J}\frac{\partial v}{\partial \xi} + \frac{x_\eta x_\xi}{J}\frac{\partial v}{\partial \eta}\right)\right] \\
&+ \frac{\partial}{\partial \eta}\left[\frac{\mu_t}{Re}\left[-\frac{x_\xi y_\eta}{J}\frac{\partial u}{\partial \xi} + \frac{x_\xi^2}{J}\frac{\partial v}{\partial \eta} + \frac{x_\eta y_\xi}{J}\frac{\partial u}{\partial \xi} + \frac{x_\xi y_\xi}{J}\frac{\partial v}{\partial \eta}\right)\right]
\end{aligned} \tag{2.92}$$

Transport Equation for Turbulent Kinetic Energy

The transport equation for turbulent kinetic energy k in curvilinear coordinates is obtained by substituting $\phi = k$ and the effective diffusivity $\Gamma = \frac{1+\mu_t/\sigma_k}{Re}$ into the general scalar transport equation. The resulting form reads:

$$\frac{\partial(Jk)}{\partial\tau} + \frac{\partial Uk}{\partial\xi} + \frac{\partial Vk}{\partial\eta} = \frac{\partial}{\partial\xi}\left[\frac{1+\mu_t/\sigma_k}{Re}\left(\frac{c_1}{J}\frac{\partial k}{\partial\xi} - \frac{c_2}{J}\frac{\partial k}{\partial\eta}\right)\right] + \frac{\partial}{\partial\eta}\left[\frac{1+\mu_t/\sigma_k}{Re}\left(-\frac{c_2}{J}\frac{\partial k}{\partial\xi} + \frac{c_3}{J}\frac{\partial k}{\partial\eta}\right)\right] + \frac{J\mu_t}{Re}G - JCD^{\varepsilon} \tag{2.93}$$

where G denotes the production of turbulent kinetic energy, and the last term accounts for its dissipation via ε, with model constant C_D.

The transport equation for the dissipation rate of turbulent kinetic energy ε is formulated using $\phi = \varepsilon$ and $\Gamma = \frac{1+\mu_t/\sigma_\varepsilon}{Re}$. The equation takes the form:

$$\frac{\partial(J\varepsilon)}{\partial\tau} + \frac{\partial U\varepsilon}{\partial\xi} + \frac{\partial(V\varepsilon)}{\partial\eta} = \frac{\partial}{\partial\xi}\left[\frac{1+\mu_t I\sigma\varepsilon}{Re}\left(\frac{c_1}{J}\frac{\partial_\varepsilon}{\partial\xi} - \frac{c_2}{J}\frac{\partial\varepsilon}{\partial}\right)\right] + \frac{\partial}{\partial\eta}\left[\frac{1+\mu_t I\sigma_\varepsilon}{Re}\left(-\frac{c_2}{J}\frac{\partial\varepsilon}{\partial\xi} + \frac{c_3}{J}\frac{\partial\varepsilon}{\partial\eta}\right)\right] + \frac{C_1 J\mu_t}{Re}\frac{G\varepsilon}{k} - JC_2\frac{\varepsilon^2}{k} \tag{2.94}$$

In this formulation, C_1 and C_2 are model constants, and the source terms represent production and destruction of dissipation rate due to turbulence.

The convection–diffusion equation for temperature θ is obtained by substituting $\phi = \theta$ and an effective thermal diffusivity of

$$\Gamma = \frac{1}{Re \cdot Pr} + \frac{\mu_t}{Re \cdot Pr_t} \tag{2.95}$$

into the general form. The resulting energy equation is:

$$\frac{\partial(J\theta)}{\partial\tau} + \frac{\partial U\theta}{\partial\xi} + \frac{\partial V\theta}{\partial\eta} = \frac{\partial}{\partial\xi}\left[\left(\frac{1}{RePr} + \frac{\mu_t}{RePr_t}\right)\left(\frac{c_1}{J}\frac{\partial\theta}{\partial\xi} - \frac{c_2}{J}\frac{\partial\theta}{\partial\eta}\right)\right] + \frac{\partial}{\partial\eta}\left[\left(\frac{1}{RePr} + \frac{\mu_t}{RePr_t}\right)\left(-\frac{c_2}{J}\frac{\partial\theta}{\partial\xi} - \frac{c_3}{J}\frac{\partial\theta}{\partial\eta}\right)\right] \tag{2.96}$$

This expression accounts for both molecular and turbulent heat conduction in general curvilinear coordinates.

The pure conduction energy equation in a medium with thermal conductivity k_r, specific heat C_{P_r}, and density ρ_r, is obtained by choosing $\phi = \theta_r$ and $\Gamma = \frac{k_r}{Re \cdot Pr}$. The governing equation becomes:

$$\rho_r C_{P_r} \frac{\partial (J\theta_r)}{\partial \tau} = \frac{1}{Re \cdot Pr} \left\{ \frac{\partial}{\partial \xi} \left(\frac{k_r c_1}{J} \frac{\partial \theta_r}{\partial \xi} - \frac{k_r c_2}{J} \frac{\partial \theta_r}{\partial \eta} \right) + \frac{\partial}{\partial \eta} \left(- \frac{k_r c_2}{J} \frac{\partial \theta_r}{\partial \xi} + \frac{k_r c_3}{J} \frac{\partial \theta_r}{\partial \eta} \right) \right\} \quad (2.97)$$

This formulation is applicable for modeling heat conduction in solids or in low-speed fluid flows dominated by thermal diffusion.

2.6 Summary

This chapter introduced the governing equations required to simulate turbulent wind flow and heat transfer around building roofs, using a curvilinear coordinate framework. The analysis was based on the assumption of a steady, two-dimensional, incompressible, turbulent flow of a Newtonian fluid with constant properties. The derivation began with the transformation of the fundamental conservation equations—including mass, momentum, energy, and turbulence kinetic energy—into a generalized non-orthogonal coordinate system, enabling the modeling of geometrically complex building envelopes. The RNG K–ε turbulence model was incorporated to close the Reynolds-averaged Navier–Stokes equations, ensuring appropriate treatment of turbulent fluctuations and energy dissipation in the atmospheric boundary layer.

Heat transfer mechanisms including conductive heat transfer through the roof, forced convective exchange between the roof surface and ambient wind, and radiative exchange with both the sky and solar input were formulated in detail. The dimensionless forms of the governing equations were also presented, providing a scalable and general framework for application to diverse urban configurations.

This chapter establishes the mathematical and physical basis for the subsequent numerical modeling of building–wind–heat interactions. The formulations developed herein will be implemented and validated in Chap. 3 through computational simulations under realistic meteorological and architectural boundary conditions.

References

1. Lecheler, S. (2022). *Computational fluid dynamics*. Springer Fachmedien Wiesbaden. https://doi.org/10.1007/978-3-658-38453-1
2. Jayanti, S. (2018). *Computational fluid dynamics for engineers and scientists*. Springer. https://doi.org/10.1007/978-94-024-1217-8
3. Wendt, J. F. (2009). *Computational fluid dynamics*. (J. F. Wendt, Ed.). Springer. https://doi.org/10.1007/978-3-540-85056-4
4. Pepper, D. W., Pirbastami, S., & Carrington, D. B. (2020). A comparison between FEM and FVM via the method of Weighted residuals. In *50 years of CFD in engineering sciences* (pp. 753–777). Springer. https://doi.org/10.1007/978-981-15-2670-1_21

5. Daneshyar, M. (1978). Solar radiation statistics for Iran. *Solar Energy, 21*(4), 345–349. https://doi.org/10.1016/0038-092X(78)90013-0
6. Hottel, H. C. (1976). A simple model for estimating the transmittance of direct solar radiation through clear atmospheres. *Solar Energy, 18*(2), 129–134. https://doi.org/10.1016/0038-092X(76)90045-1

Chapter 3
Numerical Solution Procedures for the Governing Equations in CFD

Abstract This chapter presents a comprehensive numerical framework for solving the governing equations of turbulent flow and heat transfer in a curvilinear coordinate system using the finite volume method (FVM). Following the derivation of the generalized transport equation, the discretization procedures for the momentum, turbulence, and thermal energy equations are developed over staggered control volumes. The treatment includes steady-state formulations for velocity and turbulence quantities based on the RNG k–ε model, and transient formulations for convective and conductive heat transfer. Power-law interpolation is employed to handle convection–diffusion coupling, and linearization techniques are applied to source terms to ensure algebraic solvability. The SIMPLE algorithm is adopted for pressure–velocity coupling, and under-relaxation strategies are incorporated for numerical stability. The discretized equations are solved iteratively using the TDMA (Tri-Diagonal Matrix Algorithm). Special attention is given to near-wall modeling via wall function approaches to accurately capture heat and momentum exchange at boundaries. The chapter concludes with the integration of the thermal sub-model, which employs a coupled convective-conductive energy balance for evaluating unsteady heat transfer through building envelopes under diurnal conditions.

Keywords Finite Volume Method (FVM) · Curvilinear coordinate transformation · Staggered grid arrangement · Discretization of governing equations · Power-law scheme

3.1 Introduction

The governing equations of fluid flow and heat transfer, which comprise the Navier–Stokes equations, the energy equation, and turbulence transport models, are inherently nonlinear, coupled, and often formulated in curvilinear coordinates for complex geometries [1, 2]. Due to the absence of analytical solutions for most practical problems, these equations must be solved numerically using appropriate discretization techniques.

J. Shaeri and A. Cheshmehzangi, *Applied CFD*, Urban Sustainability,
https://doi.org/10.1007/978-981-95-8505-2_3

This chapter presents a systematic formulation of the finite volume method (FVM) for solving the governing equations in curvilinear coordinates. The discretization is performed over staggered control volumes to ensure numerical stability and consistency between pressure and velocity fields. The treatment of diffusive and convective terms, source term linearization, and boundary condition implementation are addressed in detail for each equation.

The methodology is applied to both the primitive variable equations (continuity, momentum, energy) and turbulence model equations, including the transport equations for turbulent kinetic energy kkk and dissipation rate ε. Special attention is given to the transformation of governing equations into computational space and the role of Jacobian metrics and contravariant velocity components in maintaining coordinate invariance and conservation properties.

This chapter establishes the computational foundation necessary for the numerical simulation of turbulent flow and heat transfer in complex domains, which will be further explored in the subsequent chapter with experimental validation and practical applications.

3.2 Integration of the Governing Differential Equations

The governing equations for continuity, momentum, turbulent kinetic energy, turbulence dissipation rate, and convective heat transfer (energy equation) in a curvilinear coordinate system can be expressed in a unified generalized form as follows:

$$\begin{aligned} &\frac{\partial(J\phi)}{\partial\tau} + \frac{\partial}{\partial\xi}\left(U\phi - \frac{\Gamma c_1}{J}\frac{\partial\phi}{\partial\xi}\right) + \frac{\partial}{\partial\eta}\left(V\phi - \frac{\Gamma c_3}{J}\frac{\partial\phi}{\partial\eta}\right) \\ &= -\frac{\partial}{\partial\xi}\left(\frac{\Gamma c_2}{J}\frac{\partial\phi}{\partial\eta}\right) - \frac{\partial}{\partial\eta}\left(\frac{\Gamma c_2}{J}\frac{\partial\phi}{\partial\xi}\right) + JS \end{aligned} \tag{3.1}$$

In this expression, ϕ represents the dependent variable (i.e., the transported quantity), Γ denotes the effective diffusion coefficient, and S is the source term which accounts for body forces, production and dissipation mechanisms, or other physical contributions depending on the equation.

Based on the definitions provided in Chap. 2, the specific values of ϕ, Γ, and S for the governing equations describing turbulent flow using the RNG $k-\varepsilon$ model, as well as energy transport, are summarized in Table 3.1.

By integrating Eq. (3.1) over the control volume hatched in Fig. 3.1, one obtains:

$$\begin{aligned} &\int\int \frac{\partial(J\phi)}{\partial\tau} dvd\tau + \int\int \frac{\partial}{\partial\xi}\left[U\phi - \frac{\Gamma c_1}{J}\frac{\partial\phi}{\partial\xi}\right]dvd\tau + \int\int \frac{\partial}{\partial\eta}\left[V\phi - \frac{\Gamma c_3}{J}\frac{\partial\phi}{\partial\eta}\right]dvd\tau \\ &-\int\int \frac{\partial}{\partial\xi}\left[\frac{\Gamma c_2}{J}\frac{\partial\phi}{\partial\eta}\right]dvd\tau - \int\int \frac{\partial}{\partial\eta}\left[\frac{\Gamma c_2}{J}\frac{\partial\phi}{\partial\xi}\right]dvd\tau + \int\int S\,dvd\tau \end{aligned} \tag{3.2}$$

Table 3.1 Values of ϕ, Γ, and S in the Generalized Transport Equation (Eq. 3.1) for the RNG $k-\varepsilon$ Model

Governing Equation	$\boldsymbol{\phi}$	$\boldsymbol{\Gamma}$	$\boldsymbol{S}$
Continuity	1	0	0
Momentum in x-direction	u	$\frac{1+\mu_t}{Re}$	$-y_\eta \frac{\partial p}{\partial \xi} + y_\xi \frac{\partial p}{\partial \eta} +$
$\frac{\partial}{\partial \xi}\left[\frac{\mu_t}{ReJ}\left(y_\eta^2 \frac{\partial u}{\partial \xi} - y_\eta y_\xi \frac{\partial u}{\partial \eta} - y_\eta x_\eta \frac{\partial v}{\partial \xi} + x_\eta y_\xi \frac{\partial v}{\partial \eta}\right)\right]+$			
$\frac{\partial}{\partial \eta}\left[\frac{\mu_t}{ReJ}\left(-y_\xi y_\eta \frac{\partial u}{\partial \xi} + y_\xi^2 \frac{\partial u}{\partial \eta} + x_\xi y_\eta \frac{\partial v}{\partial \xi} - x_\xi y_\xi \frac{\partial v}{\partial \eta}\right)\right]$			
Momentum in y-direction	v	$\frac{1+\mu_t}{Re}$	$-x_\eta \frac{\partial p}{\partial \xi} + x_\xi \frac{\partial p}{\partial \eta} +$
$\frac{\partial}{\partial \xi}\left[\frac{\mu_t}{ReJ}\left(-y_\eta x_\eta \frac{\partial u}{\partial \xi} + y_\eta x_\xi \frac{\partial u}{\partial \eta} + x_\eta^2 \frac{\partial v}{\partial \xi} - x_\eta x_\xi \frac{\partial v}{\partial \eta}\right)\right]+$			
$\frac{\partial}{\partial \eta}\left[\frac{\mu_t}{ReJ}\left(-x_\xi x_\eta \frac{\partial v}{\partial \xi} + x_\xi^2 \frac{\partial v}{\partial \eta} + x_\eta y_\xi \frac{\partial u}{\partial \xi} - x_\xi y_\xi \frac{\partial u}{\partial \eta}\right)\right]$			
Turbulent kinetic energy k	k	$\frac{1+\mu_t/\sigma_k}{Re}$	$\frac{\mu_t}{Re} G - C_D \varepsilon$
Dissipation rate ε	ε	$\frac{1+\mu_t/\sigma_\varepsilon}{Re}$	$\frac{C_1^\varepsilon \mu_t}{Re} \frac{\varepsilon}{k} G - C_2^{*\varepsilon} \frac{\varepsilon^2}{k}$
Thermal energy θ	θ	$\frac{1}{Re\,Pr} + \frac{\mu_t}{Re\,Pr_t}$	0

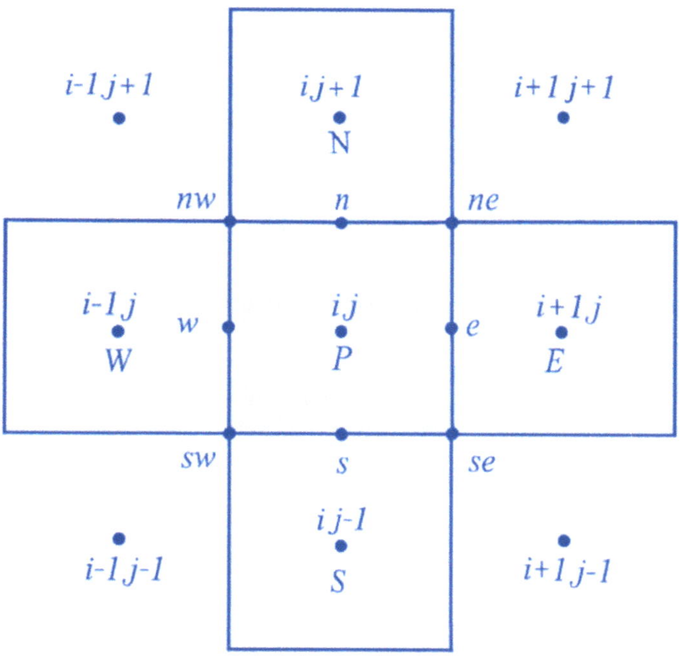

Fig. 3.1 Control volume for governing equation analysis Sa

Upon performing the integration and dividing the entire equation by $\Delta v \Delta \tau$, we obtain:

$$\begin{aligned} &\frac{(J\phi - J^0\phi^0)_P}{\Delta\tau} + \frac{1}{\Delta\xi}\left[\left(U\phi - \frac{\Gamma_{C1}}{J}\frac{\partial\phi}{\partial\xi}\right)_e - \left(U\phi - \frac{\Gamma_{C1}}{J}\frac{\partial\phi}{\partial\xi}\right)_w\right] \\ &\quad + \frac{1}{\Delta\eta}\left[\left(V\phi - \frac{\Gamma_{C3}}{J}\frac{\partial\phi}{\partial\eta}\right)_n - \left(V\phi - \frac{\Gamma_{C3}}{J}\frac{\partial\phi}{\partial\eta}\right)_s\right] \\ &= \frac{1}{\Delta\xi}\left[-\left(\frac{\Gamma_{C2}}{J}\frac{\partial\phi}{\partial\eta}\right)_e + \left(\frac{\Gamma_{C2}}{J}\frac{\partial\phi}{\partial\eta}\right)_w\right] + \frac{1}{\Delta\eta}\left[-\left(\frac{\Gamma_{C2}}{J}\frac{\partial\phi}{\partial\xi}\right)_n + \left(\frac{\Gamma_{C2}}{J}\frac{\partial\phi}{\partial\xi}\right)_s\right] + (JS)_P \end{aligned} \tag{3.3}$$

For the sake of computational convenience, it is commonly assumed that $\Delta\xi = \Delta\eta = 1$. In the left-hand side of Eq. (3.3), spatial derivatives of the scalar variable ϕ with respect to ξ and η are present. These derivatives must be approximated numerically, which necessitates an assumed distribution for ϕ.

In applying the finite volume method to transport equations, it is essential to approximate both the diffusive and convective terms in the governing differential equation. Since diffusion terms involve second-order derivatives, the values of ϕ are required at the cell centers, whereas for convection terms, the values are needed at the cell faces.

Given that the distribution of ϕ is not known a priori, the most immediate approach is to use central difference approximations for the first-order derivatives $\partial\phi/\partial\xi$ and $\partial\phi/\partial\eta$. Unfortunately, employing central differencing to approximate the terms in Eq. (3.3) often results in non-physical solutions in many practical scenarios.

To address this issue, interpolation schemes may be adopted, and such strategies have a significant impact on both convergence rate and solution accuracy. Based on Patankar's formulation, Eq. (3.3) can be reformulated as follows:

$$A_P\phi_P = A_E\phi_E + A_W\phi_W + A_N\phi_N + A_S\phi_S + S_\phi \tag{3.4}$$

The coefficients A_E, A_W, A_N, A_S, A_P in the above equation are defined according to the following relations:

$$A_E = D_e \cdot A(|Pe_e|) + max[-F_e, 0] \tag{3.5}$$

$$A_W = D_w \cdot A(|Pe_w|) + max[-F_w, 0] \tag{3.6}$$

$$A_N = D_n \cdot A(|Pe_n|) + max[-F_n, 0] \tag{3.7}$$

$$A_S = D_s \cdot A(|Pe_s|) + max[-F_s, 0] \tag{3.8}$$

$$A_P = A_E + A_W + A_N + A_S + (F_e - F_w + F_n - F_s) \tag{3.9}$$

The quantities D, F, and Pe appearing in the above equations are computed as follows:
For the east and west faces of the control volume:

$$F_{e,w} = U \cdot l_{e,w} \tag{3.10}$$

$$D_{e,w} = \frac{\Gamma c_1}{J} \cdot l_{e,w} \tag{3.11}$$

$$Pe_{e,w} = \frac{U \cdot l_{e,w}}{\frac{\Gamma c_1}{J}} \tag{3.12}$$

For the north and south faces of the control volume:

$$F_{n,s} = V \cdot l_{n,s} \tag{3.13}$$

$$D_{n,s} = \frac{\Gamma c_3}{J} \cdot l_{n,s} \tag{3.14}$$

$$Pe_{n,s} = \frac{V \cdot l_{n,s}}{\frac{\Gamma c_3}{J}} \tag{3.15}$$

Various interpolation schemes have been proposed for evaluating the function $A(|Pe|)$, which governs the balance between convection and diffusion. Table 3.2 presents several of these commonly used schemes.

In this chapter, the power-law scheme is employed to approximate the spatial derivatives of the transported scalar ϕ. A key consideration in Eq. (3.4) is the integration of the source term, which varies depending on whether the governing equation pertains to momentum, turbulent kinetic energy, or the rate of dissipation. Nevertheless, the integration of the source term can consistently be expressed in the following linearized form:

$$S_\phi = S_c + S_p \phi_p \tag{3.16}$$

Table 3.2 Interpolation schemes for approximating $A(|Pe|)$

$A(\|Pe\|)$	**Method**
1	Up wind
$\max\left[0, (1 - 0.1\|Pe\|)^5\right]$	Power law
$\max[0.1 - 0.5\|Pe\|]$	Hybrid
$\|Pe\|/\left[\exp(\|Pe\|) - 1\right]$	Exponential

In this expression, S_p is the coefficient of the unknown at the control volume center, ϕ_P, while S_c includes those source components independent of ϕ_P. This linearization ensures the preservation of the algebraic structure of Eq. (3.4).

By substituting these expressions into Eq. (3.4) and applying the power-law interpolation, the resulting general discretized formulation becomes:

$$A_{i,j,E} = \frac{1}{\Delta\xi}\left[\max(-U,0) + \frac{\Gamma c_1}{J\Delta\xi}(1-0.1|Pe|)^5\right]_e \tag{3.17}$$

$$A_{i,j,W} = \frac{1}{\Delta\xi}\left[\max(U,0) + \frac{\Gamma c_1}{J\Delta\xi}(1-0.1|Pe|)^5\right]_w \tag{3.18}$$

$$A_{i,j,N} = \frac{1}{\Delta\eta}\left[\max(-V,0) + \frac{\Gamma c_3}{J\Delta\eta}(1-0.1|Pe|)^5\right]_n \tag{3.19}$$

$$A_{i,j,S} = \frac{1}{\Delta\eta}\left[\max(V,0) + \frac{\Gamma c_3}{J\Delta\eta}(1-0.1|Pe|)^5\right]_s \tag{3.20}$$

$$A_{i,j,P} = A_E + A_W + A_N + A_S + (U_e - U_w + V_n - V_s) + \frac{J_P}{\Delta\tau} \tag{3.21}$$

The discretized form of the cross-diffusion source term is given by:

$$\begin{aligned} S_{i,j} = \tfrac{1}{\Delta\xi}[-\left(\tfrac{\Gamma c_2}{J}\right)_e \tfrac{\phi_{ne}-\phi_{se}}{\Delta\eta} + \left(\tfrac{\Gamma c_2}{J}\right)_w \tfrac{\phi_{nw}-\phi_{sw}}{\Delta\eta}] \\ +\tfrac{1}{\Delta\eta}[-\left(\tfrac{\Gamma c_2}{J}\right)_n \tfrac{\phi_{ne}-\phi_{nw}}{\Delta\xi} + \left(\tfrac{\Gamma c_2}{J}\right)_s \tfrac{\phi_{se}-\phi_{sw}}{\Delta\xi}] \\ +(JS_c)_P + \tfrac{J^0\phi^0}{\Delta\tau} \end{aligned} \tag{3.22}$$

In the above equations, the "max" function denotes the maximum value among the expressions inside the brackets, ensuring numerical stability in the upwinding scheme.

This methodology enables the integration of all differential equations introduced in Chapter 2 over their corresponding control volumes. As a result, each partial differential equation is transformed into a system of linear algebraic equations. It should be noted that the location of control volumes differs for the momentum equations (i.e., those for u and v) compared to other transported variables. The placement of control volume centers for each variable in the staggered grid arrangement is illustrated in Fig. 3.2.

In this chapter, a constant under-relaxation factor is applied during the iterative solution of the discretized differential equations. This factor is used uniformly across all iterations and for all control volumes associated with each variable. The implementation of under-relaxation is expressed as:

$$\phi_P^{(n)} = \phi_P^{(n-1)} + \alpha\left(\phi_P^* - \phi_P^{(n-1)}\right) \tag{3.23}$$

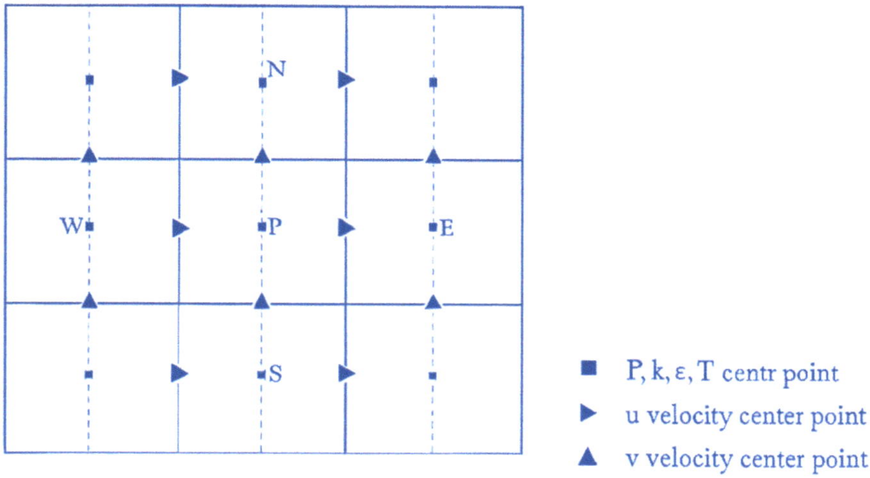

Fig. 3.2 Control Volume Centers in a Staggered Grid for Various Flow Parameters

In Eq. (3.23):

- $\phi_P^{(n)}$ is the updated value of the variable ϕ at node P in the nth iteration,
- ϕ_P^* is the newly computed value from the discretized algebraic equation,
- $\phi_P^{(n-1)}$ is the value from the previous ($(n-1)^{\text{th}}$) iteration,
- and α is the under-relaxation factor (typically $0 < \alpha \leq 1$).

Substituting the under-relaxation formulation into the general discretized equation (Eq. 3.4) yields the following modified form:

$$\frac{A_P}{\alpha}\phi_P = A_E\phi_E + A_W\phi_W + A_N\phi_N + A_S\phi_S + S_\phi + \left(\frac{1-\alpha}{\alpha}\right)A_P\phi_P^{(n-1)} \tag{3.24}$$

In this equation, α governs the degree of under-relaxation applied to the solution of ϕ, and the left-hand side coefficient matrix is adjusted accordingly to enhance numerical stability.

The coefficient A_P is computed using the following expression:

$$A_P = A_E + A_W + A_N + A_S + (F_e - F_w + F_n - F_s) + S_p \tag{3.25}$$

Here

- A_E, A_W, A_N, and A_S are the discretization coefficients for neighboring control volumes,
- F terms represent the net convective fluxes at the control volume faces, and
- S_p is the linearized source coefficient associated with ϕ_P.

3.3 Discretizaton of the x-Momentum Equation

As previously discussed, the velocity components are computed at the faces of the main control volumes. Figure 3.3 illustrates a sample staggered grid employed for solving the x-component of velocity, u.

Based on Fig. 3.3, the local Péclet numbers Pe at the four faces of the staggered control volume for the x-momentum equation are defined as follows:

$$Pe_{e_E} = \left.\frac{U}{\Gamma c_1/J}\right|_E \tag{3.26a}$$

$$Pe_{e_p} = \left.\frac{U}{\Gamma c_1/J}\right|_p \tag{3.26b}$$

$$Pe_{e_{ne}} = \left.\frac{V}{\Gamma c_3/J}\right|_{ne} \tag{3.26c}$$

$$Pe_{e_{se}} = \left.\frac{V}{\Gamma c_3/J}\right|_{se} \tag{3.26d}$$

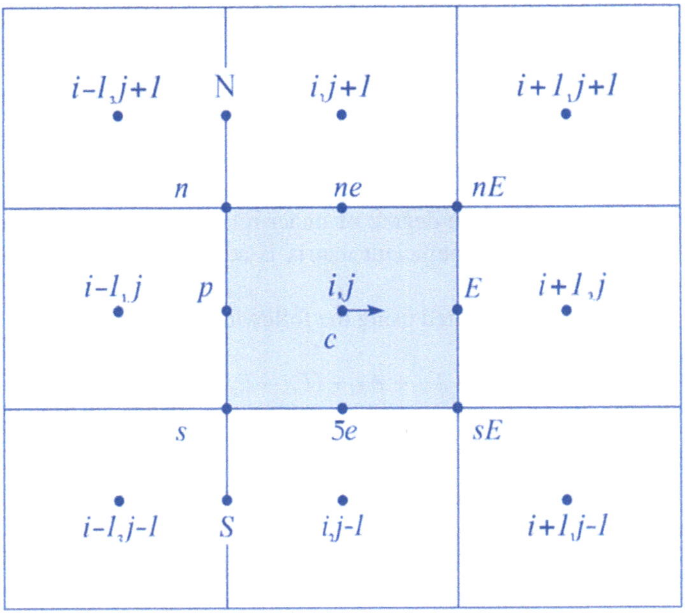

Fig. 3.3 Staggered control volume for the discretization of the ***u***-momentum equation

When applying the power-law scheme for interpolation, the coefficients $A_P - A_E$ in the steady-state x-momentum equation are computed using the following expressions:

$$A_E^u(i,j) = \frac{1}{\Delta\xi}\left[\max(-U,0) + \frac{\Gamma c_1}{J\Delta\xi}(1-0.1|Pe|)^5\right]_E \tag{3.27}$$

$$A_W^u(i,j) = \frac{1}{\Delta\xi}\left[\max(U,0) + \frac{\Gamma c_1}{J\Delta\xi}(1-0.1|Pe|)^5\right]_P \tag{3.28}$$

$$A_N^u(i,j) = \frac{1}{\Delta\eta}\left[\max(-V,0) + \frac{\Gamma c_3}{J\Delta\eta}(1-0.1|Pe|)^5\right]_{ne} \tag{3.29}$$

$$A_S^u(i,j) = \frac{1}{\Delta\eta}\left[\max(V,0) + \frac{\Gamma c_3}{J\Delta\eta}(1-0.1|Pe|)^5\right]_{se} \tag{3.30}$$

The source term, after volume integration, is expressed as:

$$\begin{aligned}
s^u(i,j) = {} & -y_{\eta_e}\frac{P_E - P_P}{\Delta\xi} + y_{\xi_e}\frac{P_{ne} - P_{se}}{\Delta\eta} + \\
& \frac{1}{\Delta\xi}\left\{-\left[\frac{1+\mu_t}{Re}\frac{c_2}{J}\frac{\partial u}{\partial\eta}\right]_E + \left[\frac{1+\mu_t}{Re}\frac{c_2}{J}\frac{\partial u}{\partial\eta}\right]_p\right\} + \\
& \frac{1}{\Delta\eta}\left\{-\left[\frac{1+\mu_t}{Re}\frac{c_3}{J}\frac{\partial u}{\partial\eta}\right]_{ne} + \left[\frac{1+\mu_t}{Re}\frac{c_2}{J}\frac{\partial u}{\partial\xi}\right]_{se}\right\} + \\
& \frac{1}{\Delta\xi}\left[\frac{\mu_t}{Re}\left(\frac{y_\eta^2}{J}\frac{\partial u}{\partial\xi} - \frac{y_\eta y_\xi}{J}\frac{\partial u}{\partial\eta} - \frac{y_\eta x_\eta}{J}\frac{\partial v}{\partial\xi} + \frac{x_\eta y_\xi}{J}\frac{\partial v}{\partial\eta}\right)\right]_E - \\
& \frac{1}{\Delta\xi}\left[\frac{\mu_t}{Re}\left(\frac{y_\eta^2}{J}\frac{\partial u}{\partial\xi} - \frac{y_\eta y_\xi}{J}\frac{\partial u}{\partial\eta} - \frac{y_\eta x_\eta}{J}\frac{\partial v}{\partial\xi} + \frac{x_\eta y_\xi}{J}\frac{\partial v}{\partial\eta}\right)\right]_P + \\
& \frac{1}{\Delta\eta}\left[\frac{\mu_t}{Re}\left(\frac{y_\xi y_\eta}{J}\frac{\partial u}{\partial\xi} + \frac{y_\xi^2}{J}\frac{\partial u}{\partial\eta} + \frac{x_\xi y_\eta}{J}\frac{\partial v}{\partial\xi} - \frac{x_\xi y_\xi}{J}\frac{\partial v}{\partial\eta}\right)\right]_{se} \\
& \frac{1}{\Delta\eta}\left[\frac{\mu_t}{Re}\left(\frac{y_\xi y_\eta}{J}\frac{\partial u}{\partial\xi} + \frac{y_\xi^2}{J}\frac{\partial u}{\partial\eta} + \frac{x_\xi y_\eta}{J}\frac{\partial v}{\partial\xi} - \frac{x_\xi y_\xi}{J}\frac{\partial v}{\partial\eta}\right)\right]_{ne}
\end{aligned} \tag{3.31}$$

By substituting Eqs. (3.27)–(3.31) into the general discretized Eq. (3.24), the x-momentum equation is fully discretized. The final step involves the application of boundary conditions.

Boundary Conditions for the x-Momentum Equation.

- No-slip condition is imposed on solid boundaries:

$$u|_{\text{wall}} = 0 \tag{3.32a}$$

- Inlet boundary condition is modeled using the following power-law velocity profile for atmospheric boundary layer flow:

$$u(y) = 1.104\, U_\infty\, y^{1/7} \tag{3.32b}$$

- Outlet boundary condition assumes vanishing velocity gradient in the streamwise direction:

$$\left.\frac{\partial u}{\partial x}\right|_{\text{outlet}} = 0 \tag{3.32c}$$

- Free-stream boundary condition specifies a uniform horizontal velocity:

$$u|_{\text{free boundary}} = 1 \tag{3.32d}$$

3.4 Discretization of the y-Momentum Equation

To compute the vertical velocity component v, the staggered grid system illustrated in Fig. 3.4 is used.

Based on this grid configuration, the local Peclet numbers at the four faces of the control volume associated with the y-momentum are expressed as:

$$\mathrm{Pe}_{ne} = \frac{U}{\Gamma\, c_1/J}|_{ne} \tag{3.33a}$$

$$\mathrm{Pe}_{nw} = \frac{U}{\Gamma\, c_1/J}|_{nw} \tag{3.33b}$$

$$\mathrm{Pe}_{N} = \frac{V}{\Gamma\, c_3/J}|_{N} \tag{3.33c}$$

$$\mathrm{Pe}_{P} = \frac{V}{\Gamma\, c_3/J}|_{P} \tag{3.33d}$$

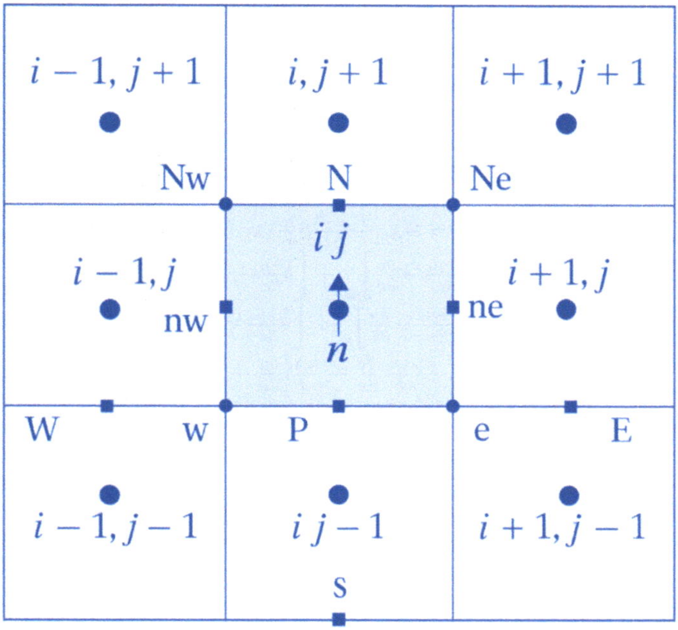

Fig. 3.4 Staggered control volume for the analysis of the ***u***-momentum equation

Assuming the use of the power-law scheme, the coefficients $A_p - A_e$ for the steady-state y-momentum equation are calculated as follows:

$$A_E^v(i,j) = \frac{1}{\Delta\xi}\left[\max(-U,0) - \frac{\Gamma c_1}{J\Delta\xi}(1 - 0.1|\text{Pe}_{ne}|)^5\right] \tag{3.34}$$

$$A_W^v(i,j) = \frac{1}{\Delta\xi}\left[\max(U,0) - \frac{\Gamma c_1}{J\Delta\xi}(1 - 0.1|\text{Pe}_{nw}|)^5\right] \tag{3.35}$$

$$A_N^v(i,j) = \frac{1}{\Delta\eta}\left[\max(-V,0) - \frac{\Gamma c_3}{J\Delta\eta}(1 - 0.1|\text{Pe}_N|)^5\right] \tag{3.36}$$

$$A_S^v(i,j) = \frac{1}{\Delta\eta}\left[\max(V,0) - \frac{\Gamma c_3}{J\Delta\eta}(1 - 0.1|\text{Pe}_P|)^5\right] \tag{3.37}$$

The central coefficient is obtained as:

$$A_P^v(i,j) = A_E^v + A_W^v + A_N^v + A_S^v + (U_{ne} - U_{nw} + U_N - U_P) \tag{3.38}$$

The discretized source term resulting from the integration is expressed as:

$$\begin{aligned} S^v(i,j) = -x_{\eta_n}\frac{P_{ne}-P_{nw}}{\Delta\xi} + x_{\xi_n}\frac{p_N-p_P}{\Delta\eta} + \\ \frac{1}{\Delta\xi}\left\{-\left[\frac{1+\mu_t}{Re}\frac{c_2}{J}\frac{\partial v}{\partial\eta}\right]_{ne} + \left[\frac{1+\mu_t}{Re}\frac{c_2}{J}\frac{\partial v}{\partial\eta}\right]_{nw}\right\} + \\ \frac{1}{\Delta\eta}\left\{-\left[\frac{1+\mu_t}{Re}\frac{c_2}{J}\frac{\partial v}{\partial\xi}\right]_{N} + \left[\frac{1+\mu_t}{Re}\frac{c_2}{J}\frac{\partial v}{\partial\xi}\right]_{P}\right\} + \\ \frac{1}{\Delta\xi}\left[\frac{\mu_t}{Re}\left(\frac{x_\eta^2}{J}\frac{\partial v}{\partial\xi} - \frac{x_\eta x_\xi}{J}\frac{\partial v}{\partial\eta} - \frac{x_\eta x_\eta}{J}\frac{\partial u}{\partial\xi} + \frac{x_\xi x_\eta}{J}\frac{\partial u}{\partial\eta}\right)\right]_{ne} - \\ \frac{1}{\Delta\xi}\left[\frac{\mu_t}{Re}\left(\frac{x_\eta^2}{J}\frac{\partial v}{\partial\xi} - \frac{x_\eta x_\xi}{J}\frac{\partial v}{\partial\eta} - \frac{y_\eta x_\eta}{J}\frac{\partial u}{\partial\xi} + \frac{x_\xi y_\eta}{J}\frac{\partial u}{\partial\eta}\right)\right]_{nw} + \\ \frac{1}{\Delta\eta}\left[\frac{\mu_t}{Re}\left(+\frac{y_\xi x_\eta}{J}\frac{\partial u}{\partial\xi} - \frac{x_\xi y_\xi}{J}\frac{\partial u}{\partial\eta} - \frac{x_\xi x_\eta}{J}\frac{\partial v}{\partial\xi} + \frac{x_\xi^2}{J}\frac{\partial v}{\partial\eta}\right)\right]_{N} - \\ \frac{1}{\Delta\eta}\left[\frac{\mu_t}{Re}\left(+\frac{y_\xi x_\eta}{J}\frac{\partial u}{\partial\xi} - \frac{x_\xi y_\xi}{J}\frac{\partial u}{\partial\eta} - \frac{x_\xi x_\eta}{J}\frac{\partial v}{\partial\xi} + \frac{x_\xi^2}{J}\frac{\partial v}{\partial\eta}\right)\right]_{P} \end{aligned} \tag{3.39}$$

3.5 Boundary Conditions for the y-Momentum Equation

The following boundary conditions are imposed for the y-component of velocity:

- No-slip condition on solid walls:

$$v|_{\text{wall}} = 0 \tag{3.40a}$$

- At the inlet, vertical velocity is assumed zero:

$$v(y) = 0 \tag{3.40b}$$

- At the outlet, a zero-gradient condition is applied:

$$\frac{\partial v}{\partial x}|_{\text{outlet}} = 0 \tag{3.40c}$$

- At the free boundary, the vertical gradient in the y-direction is set to zero:

$$\frac{\partial v}{\partial y}|_{\text{free}} = 0 \tag{3.40d}$$

3.6 Pressure Correction Equation

During the discretization of the momentum equations, the pressure field is initially assumed and subsequently corrected [3, 4]. The goal of this correction is to ensure that the velocity field obtained from the momentum equations satisfies the continuity equation as well. To perform the pressure correction, the control volume shown in Fig. 3.5 is considered.

Based on the guessed pressure field, the momentum equations in the x- and y-directions are expressed, in accordance with the general transport equation (Eq. 3.4), as follows:

$$A_p^u u_p = A_E^u u_E + A_W^u u_W + A_N^u u_N + A_S^u u_S + S^u - \frac{y_{\eta_e}}{\Delta\xi}(p_E - p_P) + \frac{y_{\xi_e}}{\Delta\eta}(p_{ne} - p_{se}) \tag{3.41}$$

$$A_p^v v_p = A_E^v v_E + A_W^v v_W + A_N^v v_N + A_S^v v_S + S^v + \frac{x_{\eta_n}}{\Delta\xi}(p_{ne} - p_{nw}) - \frac{x_{\xi_n}}{\Delta\eta}(p_N - p_P) \tag{3.42}$$

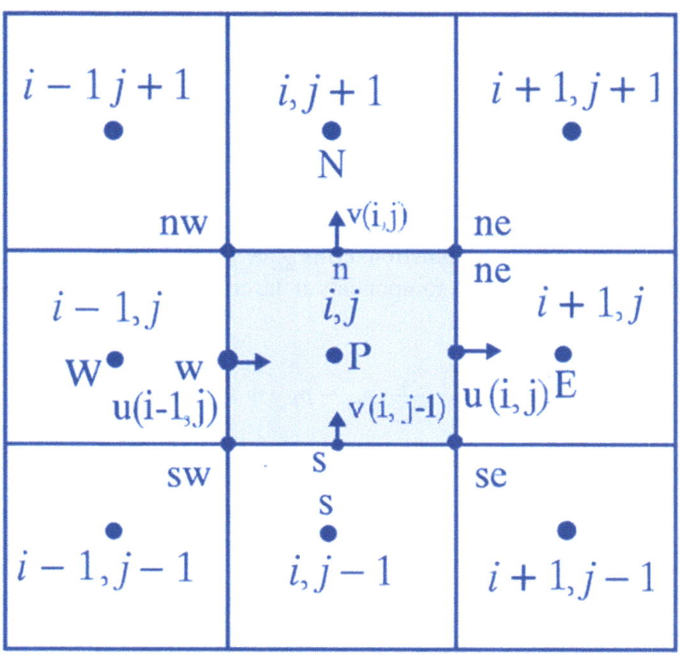

Fig. 3.5 Staggered control volume for the discretization of the pressure equation

Following the pressure correction, the corrected pressure field is denoted by P^*. Accordingly, based on Eq. (3.4), the momentum equations are reformulated as:

$$A_p^u u_p^* = A_E^u u_E^* + A_W^u u_W^* + A_N^u u_N^* + A_S^u u_S^* + S^u - \frac{y_{\eta_e}}{\Delta \xi}(p_E^* - p_P^*) + \frac{y_{\xi_e}}{\Delta \eta}(p_{ne}^* - p_{se}^*) \tag{3.43}$$

$$A_p^v v_p^* = A_E^v v_E^* + A_W^v v_W^* + A_N^v v_N^* + A_S^v v_S^* + S^v + \frac{y_{\eta_n}}{\Delta \xi}(p_{ne}^* - p_{nw}^*) - \frac{y_{\xi_n}}{\Delta \eta}(p_N^* - p_P^*) \tag{3.44}$$

By subtracting Eq. (3.41) from Eq. (3.43), and Eq. (3.42) from Eq. (3.44), the following correction equations for the velocity field are obtained:

$$A_p^u u_{P\prime} = A_E^u u_{E\prime} + A_W^u u_{W\prime} + A_N^u u_{N\prime} + A_S^u u_{S\prime} - \frac{y_{\eta_e}}{\Delta \xi}(p_{E\prime} - p_{P\prime}) + \frac{y_{\xi_e}}{\Delta \eta}(p_{ne\prime} - p_{se\prime}) \tag{3.45}$$

$$A_p^v v_{P\prime} = A_E^v v_{E\prime} + A_W^v v_{W\prime} + A_N^v v_{N\prime} + A_S^v v_{S\prime} + \frac{y_{\eta_n}}{\Delta \xi}(p_{ne\prime} - p_{nw\prime}) - \frac{y_{\xi_n}}{\Delta \eta}(p_{N\prime} - p_{P\prime}) \tag{3.46}$$

Here, the correction terms $u\prime$, $v\prime$, and $p\prime$ are defined as:

$$u\prime = u^* - u \tag{3.47}$$

$$v\prime = v^* - v \tag{3.48}$$

$$p\prime = p^* - p \tag{3.49}$$

In the above equations, the summation terms $\sum A_a^u u_{a\prime}$ and $\sum A_a^v v_{a\prime}$ are neglected.

Thus, the corrected velocity components at the control volume center are updated as:

$$u_P^* = u_p + \frac{1}{A_p^u}\left[-\frac{y_{\eta_e}}{\Delta \xi}(p_{E\prime} - p_{P\prime}) + \frac{y_{\xi_e}}{\Delta \eta}(p_{ne\prime} - p_{se\prime})\right] \tag{3.50}$$

$$v_P^* = v_p + \frac{1}{A_p^v}\left[\frac{y_{\eta_n}}{\Delta \xi}(p_{ne\prime} - p_{nw\prime}) - \frac{y_{\xi_n}}{\Delta \eta}(p_{N\prime} - p_{P\prime})\right] \tag{3.51}$$

By substituting the above correction relations into the continuity equation over the control volume depicted in Figs. 3.3, 3.3.4, 3. 5, the pressure correction equation is obtained as follows:

$$\begin{aligned}
&-\left(\frac{y_{\eta_e}}{\Delta\xi}\right)^2\frac{1}{A_P^u(i,j)}(p_{E'}-p_{P'})+\left(\frac{y_{\eta_w}}{\Delta\xi}\right)^2\frac{1}{A_P^u(i-1,j)}(p_{P'}-p_{W'})\\
&-\left(\frac{y_{\xi_e}}{\Delta\eta}\right)^2\frac{1}{A_P^v(i,j)}(p_{N'}-p_{P'})+\left(\frac{y_{\xi_s}}{\Delta\eta}\right)^2\frac{1}{A_P^v(i,j-1)}(p_{P'}-p_{S'})\\
&\quad=\left(\frac{x_{\eta_e}}{\Delta\xi}v_e-\frac{x_{\eta_w}}{\Delta\xi}v_w+\frac{x_{\xi_n}}{\Delta\eta}u_n-\frac{x_{\xi_s}}{\Delta\eta}u_s\right)\\
&\quad-\left(\frac{y_{\eta_e}}{\Delta\xi}u_{i,j}-\frac{y_{\eta_w}}{\Delta\xi}u_{i-1,j}+\frac{x_{\xi_s}}{\Delta\eta}v_{i,j-1}-\frac{x_{\xi_e}}{\Delta\eta}v_{i,j}\right)\\
&-\left[\left(\frac{y_{\xi_e}y_{\eta_e}}{\Delta\xi\,\Delta\eta}\right)\frac{1}{A_P^u(i,j)}(p_{ne'}-p_{se'})+\left(\frac{y_{\xi_w}y_{\eta_w}}{\Delta\xi\,\Delta\eta}\right)\frac{1}{A_P^u(i-1,j)}(p_{nw'}-p_{sw'})\right.\\
&\left.-\left(\frac{y_{\xi_n}y_{\eta_n}}{\Delta\xi\,\Delta\eta}\right)\frac{1}{A_P^v(i,j)}(p_{ne'}-p_{nw'})+\left(\frac{y_{\xi_s}y_{\eta_s}}{\Delta\xi\,\Delta\eta}\right)\frac{1}{A_P^v(i,j-1)}(p_{se'}-p_{sw'})\right]
\end{aligned} \tag{3.52}$$

This relation can be rearranged into the Patankar discretization format as follows:

$$A_P^p p_{P'} = A_E^p p_{E'} + A_W^p p_{W'} + A_N^p p_{N'} + A_S^p p_{S'} + S^{p'} \tag{3.53}$$

where A_P^p and its neighboring coefficients are computed based on the convective and diffusive contributions resulting from the pressure correction equation in curvilinear coordinates.

The coefficients A_E^p through A_P^p for the pressure correction equation are computed using the discretized momentum equations in the x- and y-directions, as follows:

$$A_E^p(i,j) = \frac{1}{A_P^u(i,j)}\left(\frac{y_{\eta_e}}{\Delta\xi}\right)^2 \tag{3.54}$$

$$A_W^p(i,j) = \frac{1}{A_P^u(i-1,j)}\left(\frac{y_{\eta_w}}{\Delta\xi}\right)^2 \tag{3.55}$$

$$A_N^p(i,j) = \frac{1}{A_P^v(i,j)}\left(\frac{x_{\xi_n}}{\Delta\eta}\right)^2 \tag{3.56}$$

$$A_S^p(i,j) = \frac{1}{A_P^v(i,j-1)}\left(\frac{x_{\xi_s}}{\Delta\eta}\right)^2 \tag{3.57}$$

$$A_P^p(i,j) = A_E^p(i,j) + A_W^p(i,j) + A_N^p(i,j) + A_S^p(i,j) \tag{3.58}$$

Here, A_P^u and A_P^v denote the central coefficients obtained from the discretized momentum equations in the x- and y-directions, respectively.

Source Term in the Pressure Correction Equation:

The source term $S^{p'}(i,j)$ in the pressure correction equation is obtained from the continuity equation and the provisional velocities:

$$\begin{aligned} S^{p'}(i,j) = & \left(\frac{x_{\eta_e}}{\Delta\xi}\right)v_e - \left(\frac{x_{\eta_w}}{\Delta\xi}\right)v_w + \left(\frac{y_{\xi_n}}{\Delta\eta}\right)u_n - \left(\frac{y_{\xi_s}}{\Delta\eta}\right)u_s \\ & -\left(\frac{y_{\eta_e}}{\Delta\xi}\right)u_{i,j} + \left(\frac{y_{\eta_w}}{\Delta\xi}\right)u_{i-1,j} - \left(\frac{x_{\xi_n}}{\Delta\eta}\right)v_{i,j} + \left(\frac{x_{\xi_s}}{\Delta\eta}\right)v_{i,j-1} \\ & -\left(\frac{y_{\xi_e}y_{\eta_e}}{\Delta\xi\,\Delta\eta}\right)\frac{p_{ne'}-p_{se'}}{A_P^u(i,j)} + \left(\frac{y_{\xi_w}y_{\eta_w}}{\Delta\xi\,\Delta\eta}\right)\frac{p_{nw'}-p_{sw'}}{A_P^u(i-1,j)} \\ & -\left(\frac{x_{\xi_n}x_{\eta_n}}{\Delta\xi\,\Delta\eta}\right)\frac{p_{ne'}-p_{nw'}}{A_P^v(i,j)} + \left(\frac{x_{\xi_s}x_{\eta_s}}{\Delta\xi\,\Delta\eta}\right)\frac{p_{se'}-p_{sw'}}{A_P^v(i,j-1)} \end{aligned} \tag{3.59}$$

Velocity and Pressure Correction Equations:

After solving for the pressure correction field $p\prime$, the velocity components and pressure are updated as follows:

$$u^*(i,j) = u(i,j) + \frac{1}{A_P^u(i,j)}\left[-\frac{y_{\eta_e}}{\Delta\xi}(p_{E\prime} - p_{P\prime}) + \frac{y_{\xi_e}}{\Delta\eta}(p_{ne\prime} - p_{se\prime})\right] \tag{3.60}$$

$$v^*(i,j) = v(i,j) + \frac{1}{A_P^v(i,j)}\left[\frac{y_{\eta_n}}{\Delta\xi}(p_{ne\prime} - p_{nw\prime}) - \frac{y_{\xi_n}}{\Delta\eta}(p_{N\prime} - p_{P\prime})\right] \tag{3.61}$$

$$P^*(i,j) = P(i,j) + \alpha_p p\prime(i,j) \tag{3.62}$$

Here, α_p is the under-relaxation factor for pressure correction.

Boundary Conditions for the Pressure Correction Equation:

At boundaries where the normal component of velocity is prescribed, no correction is applied to the pressure or velocity. Instead:

- The boundary velocity is directly enforced in the continuity equation.
- The corresponding coefficient in the pressure correction equation is set to zero.
- A Neumann boundary condition is applied:

$$\left.\frac{\partial P}{\partial n}\right|_{\text{wall}} = 0 \tag{3.63}$$

In regions where the pressure is explicitly specified, pressure correction is also not required, implying:

$$p\prime(i,j) = 0 \tag{3.64}$$

3.7 Discretization of the Turbulent Kinetic Energy Equation

The control volume for the turbulent kinetic energy k, the dissipation rate ε, and the temperature θ is identical to that used for the pressure variable, as illustrated in Fig. 3.6. The local Péclet numbers at the cell faces are computed using Eqs. (3.65), where the effective diffusion coefficient is defined as $\Gamma = \frac{1+\mu_t/\sigma_k}{Re}$, according to Table 3.1:

$$\mathrm{Pe}_e = \left.\frac{U}{\Gamma\, c_1/J}\right|_e \tag{3.65a}$$

$$\mathrm{Pe}_w = \left.\frac{U}{\Gamma\, c_1/J}\right|_w \tag{3.65b}$$

$$\mathrm{Pe}_n = \left.\frac{V}{\Gamma\, c_3/J}\right|_n \tag{3.65c}$$

$$\mathrm{Pe}_s = \left.\frac{V}{\Gamma\, c_3/J}\right|_s \tag{3.65d}$$

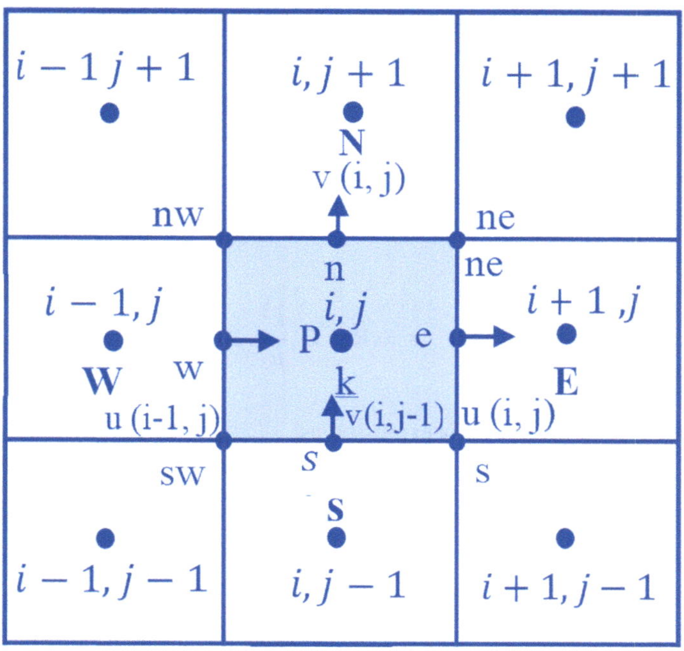

Fig. 3.6 Staggered control volume for the discretization of the turbulent kinetic energy (k) equation

Based on the power-law scheme, the coefficients A_E^k to A_P^k in the steady-state transport equation for the turbulent kinetic energy in the RNG $k-\varepsilon$ model are evaluated as follows:

$$A_E^k(i,j) = \frac{1}{\Delta\xi}\left\{|-U| - \frac{\Gamma c_1}{J\,\Delta\xi}(1-0.1|\mathrm{Pe}|)^5\right\}_e \tag{3.66}$$

$$A_W^k(i,j) = \frac{1}{\Delta\xi}\left\{|U| - \frac{\Gamma c_1}{J\,\Delta\xi}(1-0.1|\mathrm{Pe}|)^5\right\}_w \tag{3.67}$$

$$A_N^k(i,j) = \frac{1}{\Delta\eta}\left\{|-V| - \frac{\Gamma c_3}{J\,\Delta\eta}(1-0.1|\mathrm{Pe}|)^5\right\}_n \tag{3.68}$$

$$A_S^k(i,j) = \frac{1}{\Delta\eta}\left\{|V| - \frac{\Gamma c_3}{J\,\Delta\eta}(1-0.1|\mathrm{Pe}|)^5\right\}_s \tag{3.69}$$

$$A_P^k(i,j) = A_E^k + A_W^k + A_N^k + A_S^k + (U_e - U_w + V_n - V_s) \tag{3.70}$$

The source term after volume integration is expressed by:

$$\begin{aligned} S^k(i,j) = {} & \tfrac{1}{\Delta\xi}\left\{-\left[\tfrac{(1+\mu_t/\sigma_k)}{\mathrm{Re}}\tfrac{c_2}{J}\tfrac{\partial k}{\partial\eta}\right]_e + \left[\tfrac{(1+\mu_t/\sigma_k)}{\mathrm{Re}}\tfrac{c_2}{J}\tfrac{\partial k}{\partial\eta}\right]_w\right\} \\ & + \tfrac{1}{\Delta\eta}\left\{-\left[\tfrac{(1+\mu_t/\sigma_k)}{\mathrm{Re}}\tfrac{c_2}{J}\tfrac{\partial k}{\partial\xi}\right]_n + \left[\tfrac{(1+\mu_t/\sigma_k)}{\mathrm{Re}}\tfrac{c_2}{J}\tfrac{\partial k}{\partial\xi}\right]_s\right\} \\ & + \tfrac{J_p\,\mu_t(i,j)}{\mathrm{Re}}G(i,j) - J_p C_D \varepsilon(i,j) \end{aligned} \tag{3.71}$$

In the RNG $k-\varepsilon$ model, the turbulent eddy viscosity is defined as:

$$\mu_t = \mathrm{Re}\cdot C_\mu \frac{k^2}{\varepsilon} \tag{3.72}$$

Rewriting this relation for ε:

$$\varepsilon = \mathrm{Re}\cdot C_\mu \frac{k^2}{\mu_t} \tag{3.73}$$

To linearize ε in terms of k, it is expressed as:

$$\varepsilon(i,j) = Re \times C_\mu \times \frac{k^{n-1}(i,j)}{\mu_t(i,j)}k(i,j) \tag{3.74}$$

$$S_p^k = Re \times C_\mu \times \frac{k^{n-1}(i,j)}{\mu_t(i,j)} \tag{3.74a}$$

Here, $k^{(n-1)}(i,j)$ denotes the turbulent kinetic energy at control volume (i,j) from the previous iteration. Substituting Eq. (3.74) into the source term in Eq. (3.71), the source contribution is incorporated into the central coefficient of the discretized equation. The modified coefficients are then given by:

$$\begin{aligned} A_P^k(i,j) &= A_E^k(i,j) + A_W^k(i,j) + A_N^k(i,j) + A_S^k(i,j) \\ &\quad + (U_e - U_w + U_n - U_s) + J_p C_D Re C_\mu \tfrac{k^{n-1}(i,j)}{\mu_t(i,j)} \end{aligned} \tag{3.70a}$$

$$\begin{aligned} S^k(i,j) &= \tfrac{1}{\Delta\xi}\left\{-\left[\tfrac{1+\mu_t/\sigma_k}{Re}\tfrac{c_2}{J}\tfrac{\partial k}{\partial\eta}\right]_e + \left[\tfrac{1+\mu_t/\sigma_k}{Re}\tfrac{c_2}{J}\tfrac{\partial k}{\partial\eta}\right]_w\right\} + \\ &\tfrac{1}{\Delta\eta}\left\{-\left[\tfrac{1+\mu_t/\sigma_k}{Re}\tfrac{c_2}{J}\tfrac{\partial k}{\partial\xi}\right]_n + \left[\tfrac{1+\mu_t/\sigma_k}{Re}\tfrac{c_2}{J}\tfrac{\partial k}{\partial\xi}\right]_s\right\} + \tfrac{J_p\mu_t(i,j)}{Re}G(i,j) \end{aligned} \tag{3.71a}$$

Boundary Conditions for the Turbulent Kinetic Energy Equation:

At solid boundaries, the normal gradient of the turbulent kinetic energy k is assumed to vanish:

$$\left.\frac{\partial k}{\partial n}\right|_{\text{wall}} = 0 \tag{3.75a}$$

At inflow and free boundaries, a prescribed profile for k is imposed, which will be elaborated in the following chapter. At the outflow boundary, it is commonly assumed that the streamwise gradient of the turbulent kinetic energy is zero:

$$\left.\frac{\partial k}{\partial n}\right|_{\text{outlet}} = 0 \tag{3.75b}$$

In the near-wall region, the standard RNG k–ε model fails to accurately capture viscous sublayer effects due to the omission of molecular viscosity terms. Therefore, wall function formulations are employed for discretizing the turbulence energy equation near solid surfaces. Figures 3.3, 3.4, 3.5, 3.6 and 3.7 illustrates a control volume adjacent to a horizontal wall. For this cell, the coefficient A_S^k is set to:

$$A_S^k(i,j) = 0 \tag{3.75c}$$

The production term $\mu_t\left(\frac{\partial u}{\partial y}\right)^2$ in the turbulent kinetic energy equation for wall-adjacent control volumes is approximated as:

$$\mu_t\left(\frac{\partial u}{\partial y}\right)^2 = \tau_w\left(\frac{\partial u}{\partial y}\right) = \tau_w\frac{0.5(u(i,j) + u(i-1,j))}{y_P} \tag{3.75d}$$

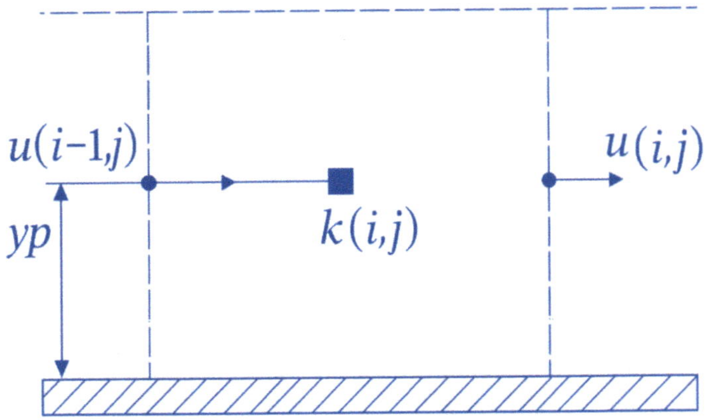

Fig. 3.7 Control volume adjacent to a Horizontal Wall for the discretization of the k-equation

Accordingly, the turbulent kinetic energy production term $G(i,j)$ for control volumes adjacent to horizontal walls is evaluated as:

$$G(i,j) = 2\left[\left(\frac{\partial u}{\partial \xi}\right)^2 + \left(\frac{\partial v}{\partial \eta}\right)^2\right] + \left(\frac{\partial v}{\partial \xi}\right)^2 + \tau_w \cdot \frac{0.5[u(i,j) + u(i-1,j)]}{y_p} \tag{3.75e}$$

he velocity gradients are approximated as:

$$\frac{\partial u}{\partial \xi} = \frac{u(i,j) - u(i-1,j)}{\Delta \xi} \tag{3.75f}$$

$$\frac{\partial v}{\partial \eta} = \frac{v(i,j) - v(i-1,j-1)}{\Delta \eta} \tag{3.75g}$$

$$\frac{\partial v}{\partial \xi} = \frac{v(i+1,j) + v(i+1,j-1) - v(i-1,j) - v(i-1,j-1)}{4\Delta \xi} \tag{3.75h}$$

The wall shear stress τ_w is determined based on the value of y^+ at the first grid point adjacent to the wall:

$$y^+ = Re(C_D C_\mu)^{0.25} k(i,j)^{0.5} y_p \tag{3.75i}$$

- If $y^+ < 11.6$:

$$\tau_w = \frac{0.5[u(i,j) + u(i-1,j)]}{\text{Re} \cdot y_p} \tag{3.75j}$$

- If $y^+ \geq 11.63$:

$$\tau_w = \frac{k\left(C_D C_\mu\right)^{0.25} k(i,j)\left[0.5(u(i,j) + u(i-1,j))\right]}{\text{In}\left(Ey^+\right)} \tag{3.75k}$$

The linearization of the dissipation term $C_D\varepsilon$ in the turbulence energy equation near horizontal walls is performed as follows:

- If $y^+ < 11.63$:

$$S_p^k(i,j) = \left(C_D C_\mu\right)^{0.75} \frac{k(i,j)^{0.5}}{y_p} y^+ \tag{3.75L}$$

- If $y^+ \geq 11.63$:

$$S_p^k(i,j) = \left(C_D\, C_\mu\right)^{0.75} \cdot \frac{\sqrt{k(i,j)}}{y_p} \cdot \frac{\ln(E \cdot y^+)}{k} \tag{3.75m}$$

Hence, for control volumes adjacent to horizontal walls, Eqs. (3.75L) and (3.75m) replace the generic source term Eq. (3.74a). Similarly, for cells adjacent to vertical walls, as illustrated in Fig. 3.8, the following relations are obtained:

$$A_E^k(i,j) = 0 \tag{3.75n}$$

$$G(i,j) = 2\left[\left(\frac{\partial u}{\partial \xi}\right)^2 + \left(\frac{\partial v}{\partial \eta}\right)^2\right] + \left(\frac{\partial u}{\partial \eta}\right)^2 + \tau_w \cdot \frac{0.5\left[v(i,j) + v(i,j-1)\right]}{x_p} \tag{3.75o}$$

The streamwise and transverse velocity gradients $\frac{\partial u}{\partial \xi}$ and $\frac{\partial v}{\partial \eta}$ are computed using Eqs. (3.75f) and (3.75g), respectively, while the cross-derivative $\frac{\partial u}{\partial \eta}$ is evaluated from:

$$\frac{du}{d\eta} = \frac{u(i,j+1) + u(i\text{-}1,j+1) - (v(i,j-1) + v(i\text{-}1,j-1))}{4\Delta\eta} \tag{3.75p}$$

For vertical wall boundaries, the non-dimensional wall distance in the x-direction, denoted x^+, is calculated as:

$$x^+ = \text{Re} \cdot (C_D\, C_\mu)^{0.25} \cdot \sqrt{k(i,j)} \cdot x_p \tag{3.75q}$$

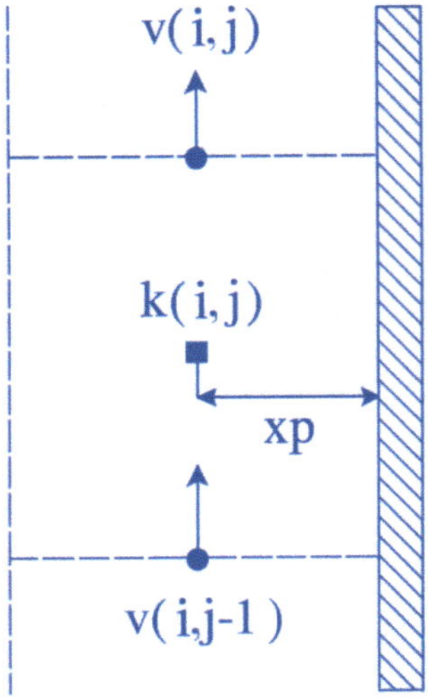

Fig. 3.8 Control volume adjacent to a vertical wall for the discretization of the k-equation

The wall shear stress τ_w is then computed depending on the range of x^+, as illustrated in Fig. 3.8:

- If $x^+ < 11.63$:

$$\tau_w = \frac{0.5(v(i,j) + v(i,j-1))}{Rex_p} \tag{3.75r}$$

- If $x^+ \geq 11.63$:

$$\tau_w = \frac{k\left(C_D C_\mu\right)^{0.25} k(i,j)\left[0.5(v(i,j) + v(i,j-1))\right]}{\text{In}(Ex^+)} \tag{3.75s}$$

The linearization procedure of the dissipation term $C_D\varepsilon$ for control volumes adjacent to vertical walls is performed as follows:

- If $x^+ < 11.63$:

$$S_p^k(i,j) = \left(C_D C_\mu\right)^{0.75} \frac{k(i,j)^{0.5}}{x_p} x^+ \tag{3.75t}$$

- If $x^+ \geq 11.63$:

$$S_p^k(i,j) = (C_D C_\mu)^{0.75} \frac{k(i,j)^{0.5}}{x_p} \frac{Ln(Ex^+)}{k} \tag{3.75u}$$

Consequently, for control volumes adjacent to vertical walls, Eqs. (3.75t) and (3.75u) replace the general source term expression given in Eq. (3.74a). In the case of curved walls, the wall function formulation remains valid, with the tangential velocity component and the local wall-normal distance used in place of horizontal or vertical components in the above relations.

3.8 Discretization of the Turbulent Dissipation Rate Equation

As previously noted, the control volume for the turbulent dissipation rate ε is defined identically to those used for pressure P and turbulent kinetic energy k, as illustrated in Fig. 3.9. Based on this control volume, the local Peclet numbers are evaluated using Eqs. (3.76). In these expressions, and in accordance with Table 3.1, the diffusion parameter is defined as $\Gamma = \frac{1+\mu_t/\sigma_\varepsilon}{Re}$.

$$\text{Pe}_e = \frac{U}{\Gamma \frac{c_1}{J}}|_e \tag{3.76a}$$

$$\text{Pe}_w = \frac{U}{\Gamma \frac{c_1}{J}}|_w \tag{3.76b}$$

$$\text{Pe}_n = \frac{V}{\Gamma \frac{c_3}{J}}|_n \tag{3.76c}$$

$$\text{Pe}_s = \frac{V}{\Gamma \frac{c_3}{J}}|_s \tag{3.76d}$$

Using the power-law scheme, the coefficients of the steady-state dissipation rate equation in the RNG $k - \varepsilon$ turbulence model are computed as follows:

$$A_E^\varepsilon(i,j) = \frac{1}{\Delta\xi}\left[\max(-U,0) - \frac{\Gamma c_1}{J\Delta\xi}(1-0.1|\text{Pe}_e|)^5\right] \tag{3.77}$$

$$A_W^\varepsilon(i,j) = \frac{1}{\Delta\xi}\left[\max(U,0) - \frac{\Gamma c_1}{J\Delta\xi}(1-0.1|\text{Pe}_w|)^5\right] \tag{3.78}$$

$$A_N^\varepsilon(i,j) = \frac{1}{\Delta\eta}\left[\max(-V,0) - \frac{\Gamma c_3}{J\Delta\eta}(1-0.1|\text{Pe}_n|)^5\right] \tag{3.79}$$

$$A_S^\varepsilon(i,j) = \frac{1}{\Delta\eta}\left[\max(V,0) - \frac{\Gamma c_3}{J\Delta\eta}(1-0.1|\text{Pe}_s|)^5\right] \tag{3.80}$$

The central coefficient is then determined from:

$$A_P^\varepsilon(i,j) = A_E^\varepsilon(i,j) + A_W^\varepsilon(i,j) + A_N^\varepsilon(i,j) + A_S^\varepsilon(i,j) + (U_e - U_w + U_n - U_s) \tag{3.81}$$

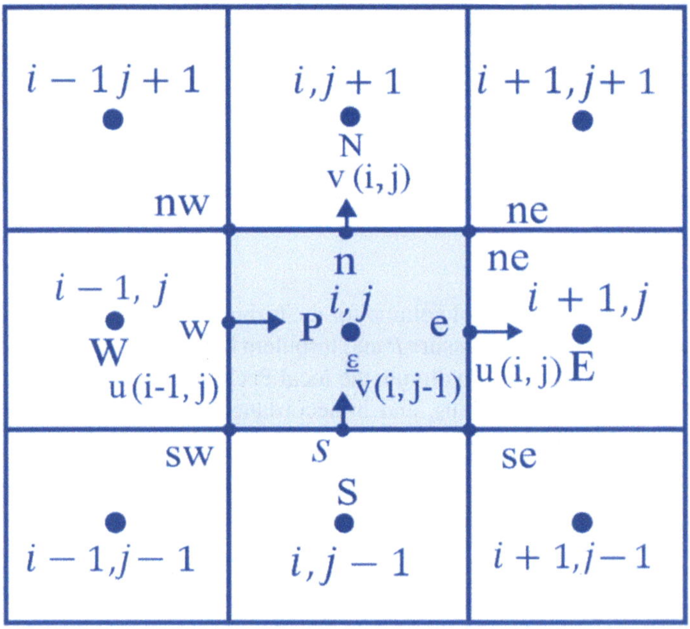

Fig. 3.9 Staggered control volume for the discretization of the ε-equation

The source term is derived via integration and expressed as:

$$\begin{aligned} S^{\varepsilon}(i,j) = {} & \tfrac{1}{\Delta\xi}\left\{-\left[\tfrac{(1+\mu_t/\sigma_\varepsilon)}{Re}\tfrac{c_2}{J}\tfrac{\partial k}{\partial\eta}\right]_e + \left[\tfrac{(1+\mu_t/\sigma_\varepsilon)}{Re}\tfrac{c_2}{J}\tfrac{\partial k}{\partial\eta}\right]_w\right\} \\ & + \tfrac{1}{\Delta\eta}\left\{-\left[\tfrac{(1+\mu_t/\sigma_\varepsilon)}{Re}\tfrac{c_2}{J}\tfrac{\partial k}{\partial\xi}\right]_n + \left[\tfrac{(1+\mu_t/\sigma_\varepsilon)}{Re}\tfrac{c_2}{J}\tfrac{\partial k}{\partial\xi}\right]_s\right\} \\ & + J_p C_{\varepsilon 1}\tfrac{\mu_t}{Re}\tfrac{\varepsilon}{k}G(i,j) - J_p C^*_{2\varepsilon}\tfrac{\varepsilon^2}{k} \end{aligned} \tag{3.82}$$

To linearize the final term in Eq. (3.83), the following transformation is applied:

$$J_p C^*_{2\varepsilon}\frac{\varepsilon^2(i,j)}{k(i,j)} = J_p C^*_{2\varepsilon}\frac{\varepsilon^{n-1}(i,j)}{k(i,j)}\varepsilon^n(i,j) \tag{3.83}$$

Accordingly, the source term becomes a linear function of ε^n, and the central coefficient in the discretized equation is modified to include this contribution:

$$\begin{aligned} A^{\varepsilon}_P(i,j) = {} & A^{\varepsilon}_E(i,j) + A^{\varepsilon}_W(i,j) + A^{\varepsilon}_N(i,j) + A^{\varepsilon}_S(i,j) + (U_e - U_w + U_n - U_s) \\ & + J_p C^*_{2\varepsilon}\tfrac{\varepsilon^{n-1}(i,j)}{k(i,j)} \end{aligned} \tag{3.84}$$

Moreover, substituting the expression for ε from Eq. (3.74) into the production term, the full source term is rewritten as:

$$S^{\varepsilon}(i,j) = \frac{1}{\Delta\xi}\left\{-\left[\frac{(1+\mu_t/\sigma_\varepsilon)}{Re}\frac{c_2}{J}\frac{\partial k}{\partial\eta}\right]_e + \left[\frac{(1+\mu_t/\sigma_\varepsilon)}{Re}\frac{c_2}{J}\frac{\partial k}{\partial\eta}\right]_w\right\}$$
$$+\frac{1}{\Delta\eta}\left\{-\left[\frac{(1+\mu_t/\sigma_\varepsilon)}{Re}\frac{c_2}{J}\frac{\partial k}{\partial\xi}\right]_n + \left[\frac{(1+\mu_t/\sigma_\varepsilon)}{Re}\frac{c_2}{J}\frac{\partial k}{\partial\xi}\right]_s\right\} \quad (3.85)$$
$$+J_p C_{\varepsilon 1} C_\mu k(i,j) G(i,j)$$

Boundary Conditions for the Turbulent Dissipation Rate Equation:

At the inlet and free boundaries, a prescribed distribution for the turbulent dissipation rate, ε, is assumed, which will be detailed in the subsequent chapter. At the outlet boundary, the streamwise gradient of ε is assumed to be zero, implying no variation of the dissipation rate in the x-direction:

$$\left.\frac{\partial\varepsilon}{\partial x}\right|_{\text{outlet}} = 0 \quad (3.86)$$

No specific boundary condition is imposed directly on solid walls. However, for the first control volume adjacent to horizontal and vertical walls, appropriate values of ε are assigned based on the following empirical wall-function expressions.

For a control volume adjacent to a horizontal wall, as illustrated in Fig. 3.7, the turbulent dissipation rate is computed as:

$$\varepsilon(i,j) = \frac{C_\mu^{3/4}}{C_D^{1/4}}\frac{k(i,j)^{3/2}}{\kappa y_p} \quad (3.87)$$

Similarly, for vertical walls, referring to Fig. 3.8, the expression becomes:

$$\varepsilon(i,j) = \frac{C_\mu^{3/4}}{C_D^{1/4}}\frac{k(i,j)^{3/2}}{\kappa x_p} \quad (3.88)$$

For curved wall boundaries, the above relations are employed by substituting the tangential velocity and the normal distance from the control volume center to the wall.

3.9 Discretization of the Turbulent Viscosity Equation

At the end of each iteration, the eddy viscosity μ_t is evaluated at the center of each control volume based on the updated values of the turbulent kinetic energy k and its dissipation rate ε using the turbulence viscosity relation:

$$\mu_t(i,j) = C_\mu Re \frac{k(i,j)^2}{\varepsilon(i,j)} \tag{3.89}$$

Since the control volumes for μ_t, k, and ε are identical in location, the above equation is directly applied as illustrated in Fig. 3.9. On solid boundaries, the turbulent viscosity is set to zero:

$$\mu_t = 0 \text{ on solid walls}$$

3.10 Numerical Solution Strategy for the Governing Equations

Each of the governing equations for fluid flow is discretized using the generalized finite volume form (see Eq. 3.25), along with the appropriate boundary conditions. This process results in a system of algebraic equations for each variable, which are solved iteratively. The procedure is repeated for all equations until convergence is achieved and the numerical solution satisfies the governing physical laws.

In this study, the SIMPLE (Semi-Implicit Method for Pressure-Linked Equations) algorithm is adopted to solve the coupled momentum and pressure equations. The solution process proceeds through the following steps:

1. Initialization of all flow variables.
2. Solution of the discretized momentum equations in the x- and y-directions.
3. Solution of the pressure correction equation to update the velocity and pressure fields.
4. Solution of the turbulent kinetic energy equation using the corrected velocity field.
5. Solution of the turbulent dissipation rate equation also based on the updated velocity field.
6. Computation of the eddy viscosity μ_t from the updated values of k and ε.
7. Return to Step 2, and repeat the process until convergence criteria are satisfied.

The convergence criterion employed in this work is based on the global residual of each transport equation across all control volumes. A residual threshold of 10^{-4} is used for all flow equations.

Each system of algebraic equations resulting from discretization is solved using the Tri-Diagonal Matrix Algorithm (TDMA), also referred to as the line-by-line iterative method. Once the flow field is fully converged, the results are utilized as inputs for solving the heat transfer equations.

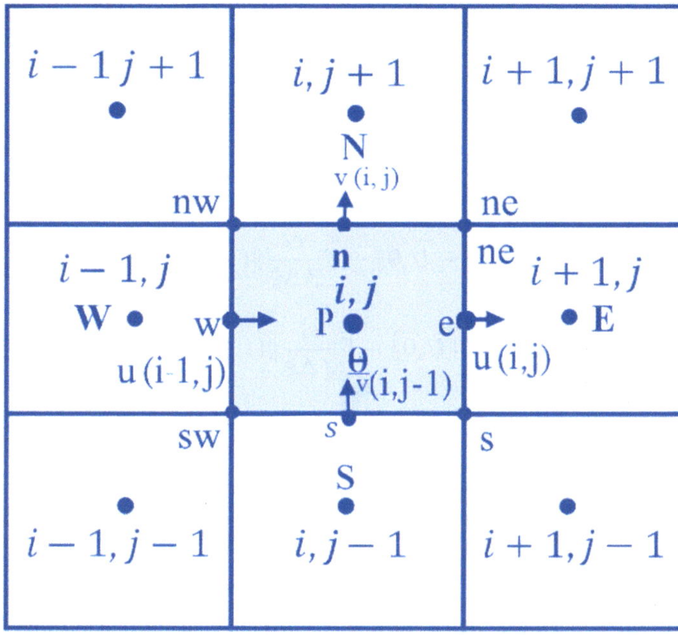

Fig. 3.10 Staggered control volume for the discretization of the energy equation

3.11 Energy Equation (Convective Heat Transfer)

To solve the energy equation and compute the temperature field θ, a control volume similar to those used for P, k, ε, and μ_t is considered, as shown in Fig. 3.10. Based on this figure, the Peclet numbers are calculated using Eqs. (3.89). In these equations, and according to Table 3.1, the effective diffusivity is given by:

$$\Gamma = \left(\frac{1}{\mathrm{RePr}} + \frac{\mu_t}{\mathrm{RePr}_t}\right) \tag{3.90}$$

The local Peclet numbers in each direction are computed as follows:

$$Pe_e = \left.\frac{U}{\Gamma\frac{c_1}{J}}\right|_e \tag{3.91a}$$

$$Pe_w = \left.\frac{U}{\Gamma\frac{c_1}{J}}\right|_w \tag{3.91b}$$

$$Pe_n = \left.\frac{V}{\Gamma\frac{c_3}{J}}\right|_n \tag{3.91c}$$

$$Pe_s = \left.\frac{V}{\Gamma\frac{c_3}{J}}\right|_s \tag{3.91d}$$

Using the power-law scheme, the coefficients A_E^θ to A_S^θ for the transient convective heat transfer equation are evaluated as:

$$A_E^\theta(i,j) = \frac{1}{\Delta\xi}\left\{\| -U,0\| - \Gamma\frac{c_1}{J\Delta\xi}\|(1-0.1|Pe)^5,0\|\right\}_e \tag{3.92}$$

$$A_W^\theta(i,j) = \frac{1}{\Delta\xi}\left\{\|U,0\| - \Gamma\frac{c_1}{J\Delta\xi}\|(1-0.1|Pe)^5,0\|\right\}_w \tag{3.93}$$

$$A_N^\theta(i,j) = \frac{1}{\Delta\eta}\left\{\| -V,0\| - \Gamma\frac{c_3}{J\Delta\eta}\|(1-0.1|Pe)^5,0\|\right\}_n \tag{3.94}$$

$$A_S^\theta(i,j) = \frac{1}{\Delta\eta}\left\{\|V,0\| - \Gamma\frac{c_3}{J\Delta\eta}\|(1-0.1|Pe)^5,0\|\right\}_s \tag{3.95}$$

According to Eq. (3.21), the total coefficient A_P^θ in the transient heat transfer equation is computed as:

$$A_P^\theta(i,j) = a_E^\theta(i,j) + a_W^\theta(i,j) + a_N^\theta(i,j) + a_S^\theta(i,j) + (U_e - U_w + U_n - U_s) + \frac{J_p}{\Delta\tau} \tag{3.96}$$

The source term $S^\theta(i,j)$ is derived through integration as follows:

$$\begin{aligned} S^\theta(i,j) = {} & \tfrac{1}{\Delta\xi}\left\{-\left[\left(\tfrac{1}{\mathrm{RePr}} + \tfrac{\mu_t}{\mathrm{RePr}_t}\right)\tfrac{c_2}{J}\tfrac{\partial\theta}{\partial\eta}\right]_e + \left[\left(\tfrac{1}{\mathrm{RePr}} + \tfrac{\mu_t}{\mathrm{RePr}_t}\right)\tfrac{c_2}{J}\tfrac{\partial\theta}{\partial\eta}\right]_w\right\} \\ & + \tfrac{1}{\Delta\eta}\left\{-\left[\left(\tfrac{1}{\mathrm{RePr}} + \tfrac{\mu_t}{\mathrm{RePr}_t}\right)\tfrac{c_2}{J}\tfrac{\partial\theta}{\partial\xi}\right]_n + \left[\left(\tfrac{1}{\mathrm{RePr}} + \tfrac{\mu_t}{\mathrm{RePr}_t}\right)\tfrac{c_2}{J}\tfrac{\partial\theta}{\partial\xi}\right]_s\right\} \\ & + \tfrac{J_p}{\Delta\tau}\theta^\circ \end{aligned} \tag{3.97}$$

where θ° denotes the value of θ at the previous time step.

Boundary Conditions for the Convective Heat Transfer Equation:

At the inlet and open boundary, a periodic profile is prescribed for the ambient temperature to account for diurnal variations. The temperature variation is expressed as:

$$T_a = \left(\frac{T_{max} + T_{min}}{2}\right) + \left(\frac{T_{max} - T_{min}}{2}\right)\cos\left(\frac{\pi(\mathrm{t}-15)}{12}\right) \tag{3.98a}$$

In the above expression:

- $T_{\max}$ and $T_{\min}$ are the maximum and minimum ambient temperatures (in °C) during a specific day at a given location.
- t denotes time in hours.

At the outlet boundary, it is assumed that there is no axial temperature gradient, leading to the following Neumann boundary condition:

$$\frac{\partial \theta}{\partial x}|_{\text{outlet}} = 0 \tag{3.98b}$$

For the solid boundaries, which include vertical walls and the ground (excluding the roof), adiabatic boundary conditions are imposed, implying zero normal heat flux. These conditions are mathematically formulated as:

$$\frac{\partial \theta}{\partial x}|_{\text{vertical wall}} = 0 \tag{3.98c}$$

$$\frac{\partial \theta}{\partial y}|_{\text{ground}} = 0 \tag{3.98d}$$

The roof boundary condition is treated separately due to its time-varying surface temperature throughout the day. The methodology for determining the roof temperature is described in the subsequent section.

3.12 Conductive Heat Transfer Equation

By expanding the partial derivatives along the computational directions τ, ξ, and η, the governing equation for conduction heat transfer is reformulated in a discretized form suitable for finite volume analysis (Eq. 3.99). This equation is integrated over a control volume centered at point P, capturing the diffusive fluxes across the computational faces [5]. The corresponding control volume employed in the discretization is illustrated in Fig. 3.11.

$$\begin{aligned}
&\rho_r C_{p_r} \tfrac{1}{\Delta\tau}\left[J\theta_r - J\theta_r^\circ\right]_P \\
&= \tfrac{k_r}{\mathrm{Re}\cdot\mathrm{Pr}}\left\{\tfrac{1}{\Delta\xi}\left[\left(\tfrac{c_1}{J}\tfrac{\partial\theta_r}{\partial\xi} - \tfrac{c_2}{J}\tfrac{\partial\theta_r}{\partial\eta}\right)_e - \left(\tfrac{c_1}{J}\tfrac{\partial\theta_r}{\partial\xi} - \tfrac{c_2}{J}\tfrac{\partial\theta_r}{\partial\eta}\right)_w\right]\right. \\
&\left.+\tfrac{1}{\Delta\eta}\left[\left(-\tfrac{c_2}{J}\tfrac{\partial\theta_r}{\partial\xi} + \tfrac{c_3}{J}\tfrac{\partial\theta_r}{\partial\eta}\right)_n - \left(-\tfrac{c_2}{J}\tfrac{\partial\theta_r}{\partial\xi} + \tfrac{c_3}{J}\tfrac{\partial\theta_r}{\partial\eta}\right)_s\right]\right\}
\end{aligned} \tag{3.99}$$

Here, θ_r denotes the non-dimensional temperature in the solid domain, and θ_r° is its value at the previous time step. Subscripts $e, w, n,$ and s refer to the east, west, north, and south cell faces, respectively.

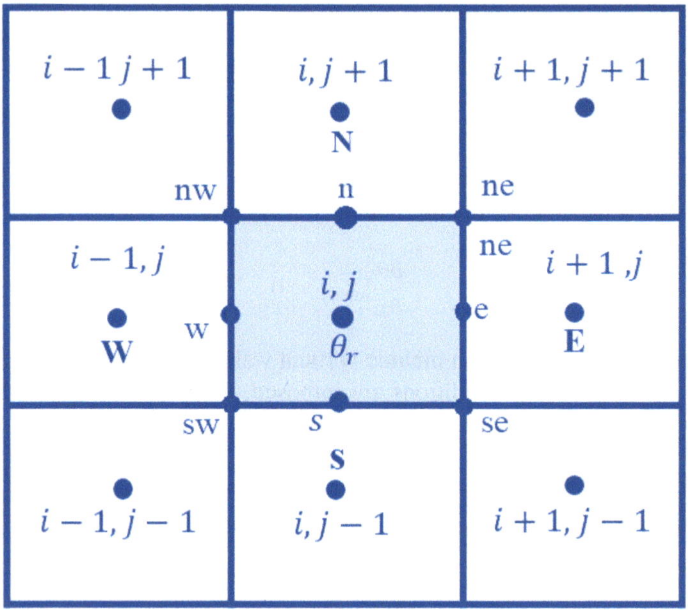

Fig. 3.11 Staggered control volume for the discretization of the conduction heat transfer equation

The partial derivatives at the control volume faces are evaluated using second-order central differences, as defined by the following expressions:

$$\left.\frac{\partial \theta_r}{\partial \xi}\right|_e = \frac{\theta_r|_E - \theta_r|_P}{\Delta \xi} \tag{3.100a}$$

$$\left.\frac{\partial \theta_r}{\partial \xi}\right|_W = \frac{\theta_r|_P - \theta_r|_W}{\Delta \xi} \tag{3.100b}$$

$$\left.\frac{\partial \theta_r}{\partial \eta}\right|_{se} = \frac{\theta_r|_{ne} - \theta_r|_{se}}{\Delta \eta} \tag{3.100c}$$

$$\left.\frac{\partial \theta_r}{\partial \eta}\right|_{sw} = \frac{\theta_r|_{nw} - \theta_r|_{sw}}{\Delta \eta} \tag{3.100d}$$

$$\left.\frac{\partial \theta_r}{\partial \zeta}\right|_s = \frac{\theta_r|_{ne} - \theta_r|_{nw}}{\Delta \xi} \tag{3.100e}$$

$$\left.\frac{\partial \theta_r}{\partial \zeta}\right|_s = \frac{\theta_r|_{se} - \theta_r|_{sw}}{\Delta \xi} \tag{3.100f}$$

$$\left.\frac{\partial \theta_r}{\partial \eta}\right|_n = \frac{\theta_r|_N - \theta_r|_p}{\Delta \eta} \tag{3.100g}$$

$$\left.\frac{\partial \theta_r}{\partial \eta}\right|_s = \frac{\theta_r|_p - \theta_r|_s}{\Delta \eta} \tag{3.100h}$$

These expressions are used to compute thermal gradients at the faces of the control volume for numerical evaluation of the heat diffusion flux.

Based on the preceding finite difference relations for temperature gradients, the discretization coefficients $A_E^{\theta_r}$ to $A_P^{\theta_r}$ required for solving the conduction heat transfer equation via the TDMA (Tri-Diagonal Matrix Algorithm) are expressed as follows:

$$A_E^{\theta_r}(i,j) = \frac{1}{\Delta \xi^2}\left(\frac{k_r}{\rho_r C_{p_r}} \cdot \frac{C_{1e}}{\mathrm{Re}\cdot \mathrm{Pr}\cdot J_e}\right) \tag{3.101}$$

$$A_W^{\theta_r}(i,j) = \frac{1}{\Delta \xi^2}\left(\frac{k_r}{\rho_r C_{p_r}} \cdot \frac{C_{1w}}{\mathrm{Re}\cdot \mathrm{Pr}\cdot J_w}\right) \tag{3.102}$$

$$A_N^{\theta_r}(i,j) = \frac{1}{\Delta \eta^2}\left(\frac{k_r}{\rho_r C_{p_r}} \cdot \frac{C_{1n}}{\mathrm{Re}\cdot \mathrm{Pr}\cdot J_n}\right) \tag{3.103}$$

$$A_S^{\theta_r}(i,j) = \frac{1}{\Delta \eta^2}\left(\frac{k_r}{\rho_r C_{p_r}} \cdot \frac{C_{1s}}{\mathrm{Re}\cdot \mathrm{Pr}\cdot J_s}\right) \tag{3.104}$$

The coefficient at the central node $A_P^{\theta_r}$ is given by the sum of the neighboring coefficients and the transient term:

$$A_P^{\theta_r}(i,j) = A_E^{\theta_r}(i,j) + A_W^{\theta_r}(i,j) + A_N^{\theta_r}(i,j) + A_S^{\theta_r}(i,j) + \frac{J_p}{\Delta \tau} \tag{3.105}$$

The source term $S^{\theta_r}(i,j)$, resulting from the integral of the right-hand side terms over the control volume, is formulated as follows:

$$\begin{aligned} S^{\theta_r}(i,j) = {} & \tfrac{1}{\Delta\xi\,\Delta\eta}\left(\tfrac{k_r}{\mathrm{Re}\cdot\mathrm{Pr}\cdot\rho_r C_{pr}}\right)\Big[\Big(-\tfrac{C_{2e}}{J_e}(\theta_{ne}-\theta_{se}) + \tfrac{C_{2w}}{J_w}(\theta_{nw}-\theta_{sw})\Big) \\ & + \Big(-\tfrac{C_{2n}}{J_n}(\theta_{ne}-\theta_{nw}) + \tfrac{C_{2s}}{J_s}(\theta_{se}-\theta_{sw})\Big)\Big] + \tfrac{J_P}{\Delta\tau}\theta_r^{\circ} \end{aligned} \tag{3.106}$$

Here, θ_r° represents the temperature value at the previous time step.

Boundary Conditions for the Heat Conduction Equation:

A portion of the roof boundary that coincides with the vertical wall is treated as thermally insulated:

$$\left.\frac{\partial \theta_r}{\partial x}\right|_{\text{vertical wall}} = 0 \tag{3.107a}$$

At the lower boundary of the roof, which is in contact with the conditioned indoor air, a constant temperature of $T = 27^{\circ}C$ is prescribed. At the upper surface of the roof, which is exposed to the outdoor environment, the following energy balance must be satisfied for each surface control volume:

$$q_s + q_{\text{cond}}|_{y\text{ex}} + q_{\text{conv}}|_{y\text{ex}} + q_{\text{sky}} = \rho_r C_{p_r} \Delta y \frac{\partial T}{\partial t} \tag{3.107b}$$

In this equation:

- q_s denotes the incident solar radiation on the surface at a specific time and location.
- q_{sky} represents the net radiative heat exchange between the roof and the sky, both of which were defined and computed in Chapter 2.
- q_{cond} corresponds to the conductive heat transfer between the upper roof layer and the layers beneath.
- q_{conv} is the convective heat exchange with the ambient outdoor air.

3.13 Calculation of Convective Heat Transfer Coefficient

The convective heat flux from the external surface of the roof to the ambient air is computed using the following relation:

$$q'' = \frac{(T_{\text{ex}} - T_p)\,\rho\, C_P\, C_\mu^{1/4}\, k_p^{1/2}}{T^+} \tag{3.108}$$

where T_{ex} denotes the temperature of the outer surface of the roof, and T_p and k_p refer to the temperature and the turbulent kinetic energy, respectively, at the first control volume above the roof surface within the computational domain.

The non-dimensional wall temperature T^+ is evaluated using the logarithmic wall function expression:

$$T^+ = Pr_t\left[\frac{1}{\kappa}\ln\left(Ey^+\right) + P\right] \tag{3.109}$$

where the correction term P is defined as:

$$P = 9.24\left[\left(\frac{Pr}{Pr_t}\right)^{3/4} - 1\right]\left[1 + 0.28\exp\left(-0.007\frac{Pr}{Pr_t}\right)\right] \tag{3.110}$$

In the above expressions:

- ρ is the air density,
- C_P is the specific heat at constant pressure,
- C_μ is the standard turbulence model constant,
- κ is the von Kármán constant,
- y^+ is the dimensionless wall distance,

- Pr and Pr_t are the molecular and turbulent Prandtl numbers, respectively, and
- E is the empirical constant in the log-law for temperature.

Once the convective heat flux q'' is determined using Eq. (3.106), the convective heat transfer coefficient h can be obtained from:

$$h = \frac{q''}{T_{\text{ex}} - T_a} \tag{3.111}$$

Here, T_a is the ambient air temperature in Kelvin, which is computed as a time-periodic function using Eq. (3.100a).

3.14 Numerical Solution Procedure for the Coupled Heat Transfer Equations

Once the steady-state flow field has been computed, the obtained velocity and turbulence parameters are employed to evaluate the temperature distribution and the convective heat transfer from the roof surface. The thermal sub-model is solved through the following sequence of steps:

1. Initialize the temperature field in the computational domain and the roof structure.
2. Solve the convective heat transfer equation within the fluid domain.
3. Solve the conductive heat transfer equation through the solid roof layers.
4. Solve the energy balance equation on the roof surface (3.107b).
5. Repeat steps 2 through 4 until thermal convergence is achieved.
6. Compute the convective heat transfer coefficient on the outer roof surface.
7. Advance the simulation by one time step and return to step 2.
8. Continue the time marching until a full 24-h simulation cycle is completed.

The convergence criterion for the thermal equations in this study is based on the magnitude of the residuals in all control volumes. Specifically, the solution is considered converged when the residuals of all discretized thermal equations fall below a threshold of 10^{-6}.

Each algebraic system resulting from the discretization of the governing equations is solved using a line-by-line iterative method based on the Tri-Diagonal Matrix Algorithm (TDMA).

The overall algorithm for solving the coupled flow and heat transfer equations can be summarized as follows:

9. Initialize all flow variables.
10. Solve the momentum equations in the x- and y-directions.

11. Solve the pressure correction equation to obtain updated pressure and velocity fields.
12. Solve the turbulent kinetic energy (k) equation using the updated velocity field.
13. Solve the turbulent dissipation rate (ε) equation using the updated velocity field.
14. Update the turbulent viscosity μ_t based on the computed k and ε.
15. Repeat steps 2 through 6 until the flow equations converge.
16. Initialize the temperature field in the fluid and solid domains.
17. Solve the convective heat transfer equation.
18. Solve the conductive heat transfer equation through the roof.
19. Solve the surface energy balance equation on the roof.
20. Repeat steps 9 through 11 until convergence of the thermal equations is achieved.
21. Compute the convective heat transfer coefficient at the roof surface.
22. Advance in time and return to step 9.
23. Complete the 24-h time cycle.

The entire simulation can be repeated over several consecutive days until a periodic and thermally stable solution is reached for both temperature distribution and heat flux.

3.15 Summary

In this chapter, a rigorous numerical methodology based on the finite volume method (FVM) has been developed for solving the governing equations of fluid flow and heat transfer in curvilinear coordinate systems. The control volume discretization strategy utilizes a staggered grid arrangement to accurately capture pressure–velocity coupling via the SIMPLE algorithm. All governing transport equations—including those for momentum, turbulent kinetic energy (k), turbulence dissipation rate (ε), and energy—are discretized using the power-law scheme, with linearized source terms and appropriate treatment of boundary conditions.

Special attention has been given to the implementation of turbulence modeling using the RNG k–ε approach, enabling the resolution of complex turbulent flow patterns with enhanced stability and accuracy. Additionally, thermal transport is modeled using both steady and transient energy equations, accounting for convection and conduction in building components. Near-wall treatments are incorporated via wall functions to address velocity and thermal gradients adjacent to solid surfaces.

The chapter establishes the computational foundation required for implementing the CFD model in practical urban or building-scale simulations. The techniques developed herein will be utilized in subsequent chapters to simulate diurnally varying external wind conditions, assess thermal comfort, and evaluate ventilation performance in urban environments.

References

1. Runchal, A. (2020). *50 Years of CFD in engineering sciences.* (A. Runchal, Ed.). Springer Singapore. https://doi.org/10.1007/978-981-15-2670-1
2. Sharma, A. (2016). Introduction to computational fluid dynamics. *International Journal of Information Science and Computing* (Vol. 3). Wiley. https://doi.org/10.1002/9781119369189
3. Zhai, Z. (John). (2020). *Computational fluid dynamics for built and natural environments.* Springer Singapore. https://doi.org/10.1007/978-981-32-9820-0
4. Blazek, J. (2015). Computational fluid dynamics: Principles and applications. *Elsevier*. https://doi.org/10.1016/C2013-0-19038-1
5. Wendt, J. F. (2009). *Computational fluid dynamics.* (J. F. Wendt, Ed.). Springer. https://doi.org/10.1007/978-3-540-85056-4

Chapter 4
Wind Tunnel Testing and Velocity Field Reconstruction for Urban and Building Applications

Abstract This chapter presents an experimental investigation of wind flow around building and urban-scale models using a wind tunnel. The primary focus is on the use of an open-circuit wind tunnel and hot-wire anemometry (HWA) for high-resolution measurements of instantaneous velocity components within the boundary layer. The chapter begins by introducing the operational principles, classification, and structural components of wind tunnels. Calibration procedures for velocity sensors, the design of vorticity generators, and surface roughness elements for replicating atmospheric boundary layer profiles are also detailed. A novel method is introduced for reconstructing two-dimensional velocity fields using single-axis hot-wire sensors, enabling accurate flow measurements without the complexity of multi-sensor arrays. Statistical and spectral analyses of the velocity signals are performed, and experimental results are compared with numerical simulations for validation. The findings support the development of reliable CFD validation techniques for assessing natural ventilation, pollutant dispersion, and wind-induced loads in climate-responsive building and urban design.

Keywords Wind Tunnel · Hot-Wire Anemometry (HWA) · Boundary Layer Profiling · CFD Validation

4.1 Introduction

In order to investigate fluid flow around architectural and urban-scale models in a wind tunnel, it is essential to first prepare all necessary experimental equipment and instrumentation [1, 2]. This includes not only selecting the appropriate measurement devices but also determining their specifications, operational principles, and associated measurement accuracies. Once the instrumentation has been properly configured, a physical model of the target structure can be fabricated with the required geometric precision, based on the specific experimental objectives and conditions.

Today, wind tunnels are widely employed at leading research institutions and universities to conduct experimental investigations of wind flow around and within

J. Shaeri and A. Cheshmehzangi, *Applied CFD*, Urban Sustainability,
https://doi.org/10.1007/978-981-95-8505-2_4

buildings and urban fabrics. These facilities enable the controlled replication of realistic flow conditions, making them indispensable tools for studying complex aerodynamic interactions in the built environment.

Given that the objective of this chapter is to examine wind flow over both individual buildings and urban configurations and considering that each wind tunnel setup—along with its associated measurement system—possesses unique operational characteristics, this chapter begins by introducing the wind tunnel facility and the instrumentation specifically designed for building and urban-scale flow analysis.

In addition to standard methodologies, this chapter introduces a novel measurement technique developed to reconstruct two-dimensional flow velocity fields using single axis hot-wire sensors. The underlying formulation, implementation procedure, and governing equations for this method are detailed in the subsequent sections.

4.2 Wind Tunnel

A wind tunnel is an experimental facility designed to generate a controlled airstream that passes over a physical model, thereby replicating real-world flow phenomena at a reduced scale [3]. Through the use of auxiliary instruments and flow conditioning systems, wind tunnels enable precise measurement and visualization of airflow characteristics around the model [2, 4]. This makes wind tunnels one of the most efficient and cost-effective tools for experimental research in aerodynamics. Figures 4.1 and 4.2 illustrate small- and large-scale open-circuit wind tunnel systems used at Tarbiat Modares University and K. N. Toosi University of Technology respectively.

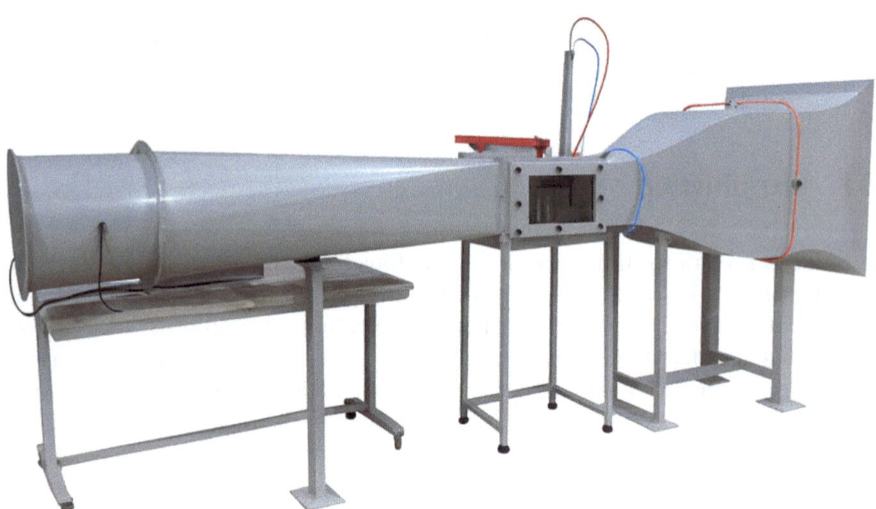

Fig. 4.1 Open-circuit wind tunnel for small-scale model testing at Tarbiat Modares University, Iran

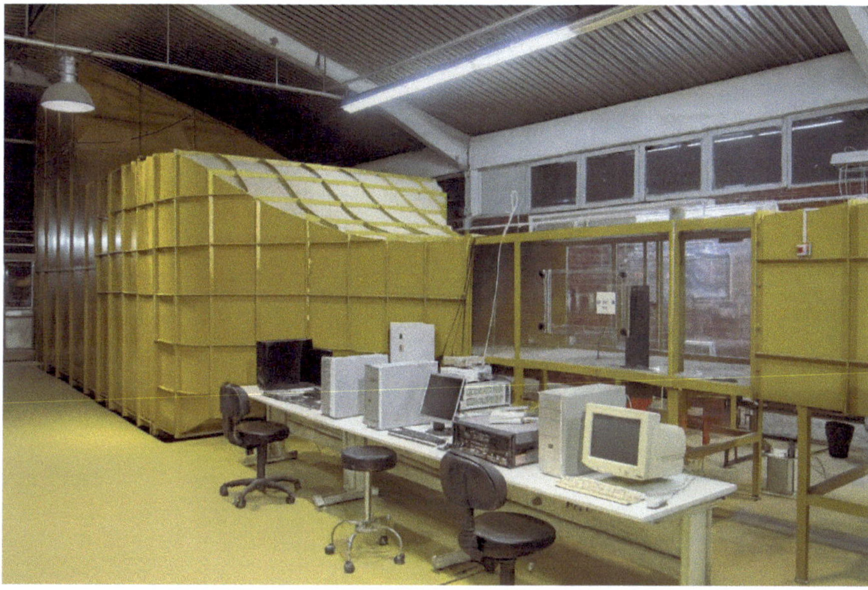

Fig. 4.2 Open-circuit wind tunnel for medium-scale model testing at K. N. Toosi University of Technology, Iran

Wind tunnels are typically classified according to several parameters, such as flow velocity, test section dimensions, and operational regime [3, 5]. From the perspective of flow compressibility, tunnels are categorized by their Mach number range as follows:

- Incompressible subsonic: $M < 0.3$,
- Compressible subsonic: $0.3 \leq M < 0.7$,
- Transonic: $0.7 \leq M \leq 1.3$,
- Supersonic: $1.3 < M \leq 5.0$, *and*
- Hypersonic: $5.0 < M \leq 10$.

Based on the target flow regime for a given study, a wind tunnel with the appropriate speed range can be selected.

Low-speed wind tunnels, which are widely used for architectural and environmental flow studies, can be further divided by flow circuit configuration into two main types:

- Open-circuit wind tunnels and
- Closed-circuit wind tunnels

Figure 4.3a illustrates the schematic and operation of a closed-circuit wind tunnel. In such tunnels as shown in Fig. 4.3b, guide vanes are used to redirect and stabilize the recirculating flow throughout the loop.

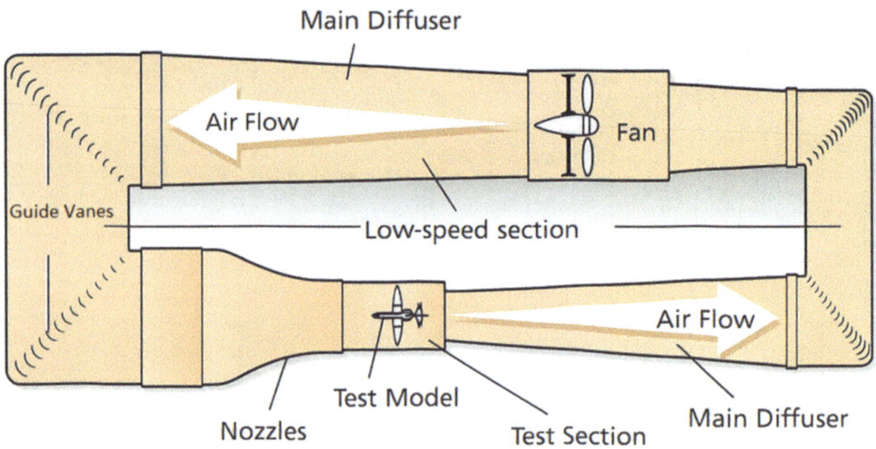

a) Operating principle of closed-circuit wind tunnels

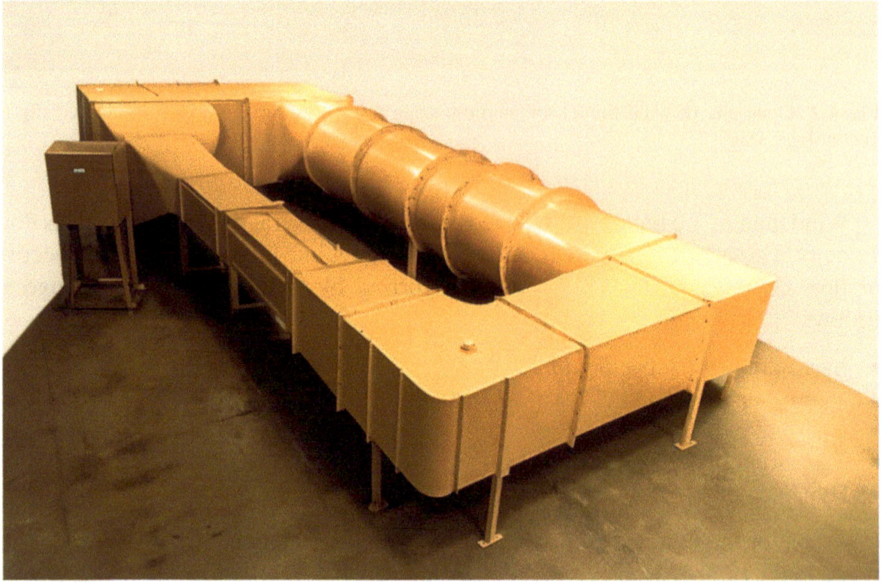

b) a case of closed-circuit wind tunnels

Fig. 4.3 Configuration and operating principles of closed-circuit Wind Tunnels. **a** Operating principle of closed-circuit wind tunnels. **b** a case of closed-circuit wind tunnels

In an open-circuit wind tunnel, ambient air is drawn into the system and, after passing through the tunnel components and test section, is expelled back into the environment [6]. The inlet and outlet flows are independent, ensuring minimal feedback effects. A typical open-circuit wind tunnel is illustrated in Fig. 4.4, where a fan draws external air into the system.

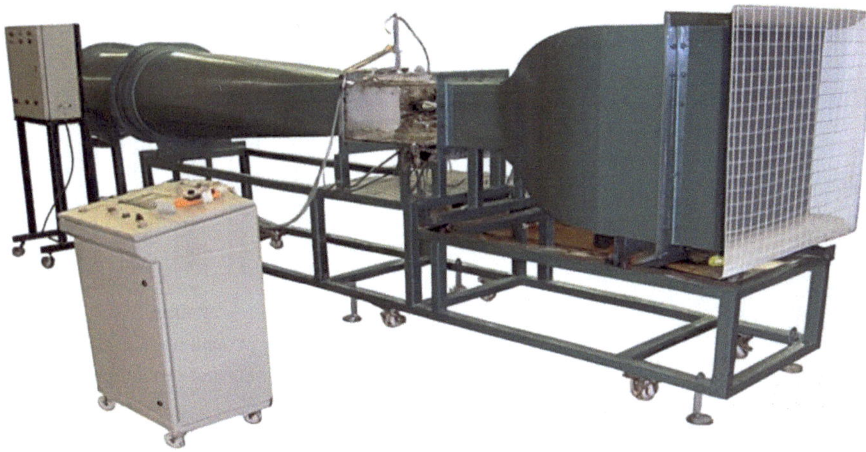

Fig. 4.4 A representative example of an open-circuit blower-type wind tunnel

Figure 4.5 illustrates the experimental wind tunnel setup employed to investigate the aerodynamic behavior of an atrium-roofed three-story building model under wind flow conditions as shown in Fig. 4.5c [7]. Figure 4.5 depicts both the major components of the open-circuit wind tunnel (see in Fig. 4.5a) and its operating principle. This experimental configuration was specifically designed to acquire high-resolution airflow velocity data (m/s), which serve as a reference for validating numerical simulation results obtained using the COMSOL Multiphysics software (Fig. 4.5b). The primary objective of this investigation was to assess the effect of wind-induced flow on the aerodynamic performance of the atrium roof structure.

The airflow first enters a wide-angle diffuser leading into a settling chamber, which often includes a honeycomb structure for flow straightening. It then passes through a contraction nozzle with a gradually decreasing cross-sectional area, accelerating the flow into the test section (Fig. 4.5d). The test section is designed to provide a uniform, steady, and predictable flow field where scaled physical models can be studied. The flow velocity in the tunnel can be adjusted by controlling the rotational speed of the driving fan.

Based on the placement of the fan relative to the test section, open-circuit wind tunnels are divided into two types:

- Blower-type (fan upstream of the test section) and
- Suction-type (fan downstream of the test section).

In a suction-type open-circuit wind tunnel with a sealed test section, the static pressure inside the chamber is lower than atmospheric pressure. As a result, air is drawn into the test section from the external environment through controlled inlet ports or compartments.

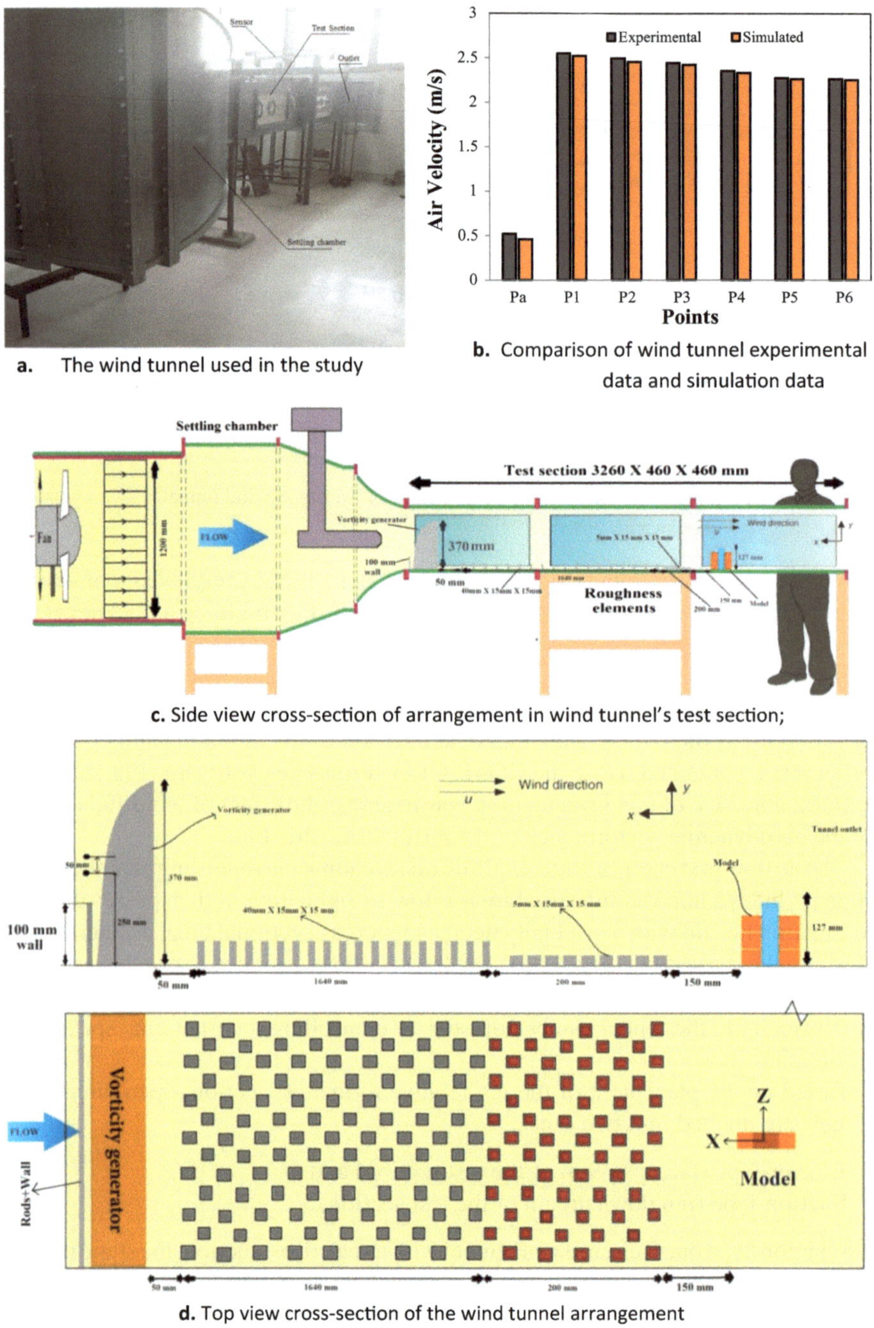

Fig. 4.5 The wind tunnel, vortex generators, and roughness elements along with barrier and rods. **a** The wind tunnel used in the study. **b** Comparison of wind tunnel experimental data and simulation data. **c.** Side view cross-section of arrangement in wind tunnel's test section. **d** Top view cross-section of the wind tunnel arrangement

4.3 Hot-Wire Anemometry

To investigate the effects of fluid flow on real environments and to develop effective flow control strategies, it is essential to accurately measure key fluid dynamic quantities. In structural engineering, for instance, understanding wind-induced loading and airflow patterns around structures—such as buildings, bridges, and stadiums—requires experimental determination of parameters such as:

- Pressure distribution,
- Velocity profiles,
- Turbulence spectrum, and
- Boundary layer thickness.

These measurements are typically carried out through wind tunnel experiments, in which a scaled physical model of the structure is tested under controlled flow conditions [8]. A critical requirement for such tests is the placement of the model within the boundary layer, and the achievement of geometric and dynamic similarity between the tunnel flow and atmospheric conditions [6]. This is accomplished by matching parameters such as the Reynolds number, reproducing realistic velocity distributions, and quantifying the turbulence spectrum.

For vibration analysis of structures exposed to wind, accurate measurement of the frequency of turbulent velocity fluctuations is particularly important. Hence, precise acquisition of flow field data is essential, as even small errors in measured quantities can lead to significant deviations in the structural design process.

One of the most influential flow properties is the instantaneous velocity vector of the fluid, which can be expressed as:

$$\vec{V}(t) = (U(t), V(t), W(t)) \tag{4.1}$$

where $U(t)$, $V(t)$, and $W(t)$ are the velocity components in the Cartesian directions x, y, and z, respectively.

Each component of instantaneous velocity can be decomposed into a mean and a fluctuating part as follows:

$$U(t) = \overline{U}(t) + u(t) \tag{4.2}$$

$$V(t) = \overline{V}(t) + v(t) \tag{4.3}$$

$$W(t) = \overline{W}(t) + w(t) \tag{4.4}$$

Accurate measurement of the turbulent components $u(t)$, $v(t)$, and $w(t)$, along with their time-domain or frequency-domain variations, is crucial for characterizing flow behavior and implementing effective control mechanisms. The frequency content of these fluctuations ranges from a few Hz in laminar flows to several kHz in highly turbulent regimes and is strongly dependent on the Reynolds number.

In experimental fluid dynamics, flow velocity can be determined by two primary approaches:

- Indirect methods, in which pressure is measured and velocity is inferred via fluid mechanics equations (e.g., using a Pitot tube and Bernoulli's law).
- Direct methods, in which velocity is measured directly using instruments such as Laser Doppler Velocimeters (LDVs) or Hot-Wire Anemometers (HWAs).

In hot-wire anemometry, the output is typically a voltage signal, which must be carefully calibrated before being interpreted. The calibration process involves establishing a relationship between voltage and velocity. Once this relationship is known, the instantaneous flow velocity can be computed using calibration equations. Hot-wire anemometry remains a powerful tool for high-resolution velocity measurements in turbulent flows, particularly when detailed temporal data is required.

In the indirect method of velocity measurement, a Pitot tube is used to determine the dynamic pressure of the flow. The mean velocity of the fluid is then calculated using fundamental principles of fluid mechanics. However, this technique is incapable of capturing turbulence fluctuations, and pressure measurement devices generally exhibit limited frequency response, typically restricting the measurement of instantaneous velocities to only a few hundred Hertz. Therefore, for applications requiring high-frequency temporal resolution or rapid response to transient flow variations, hot-wire anemometry (HWA) and laser Doppler velocimetry (LDV) are preferred.

The hot-wire anemometer is a precision instrument capable of measuring the instantaneous velocity of fluid flow with high temporal resolution. Based on these instantaneous values, additional turbulence quantities can be extracted, including:

- Mean flow velocity,
- Velocity fluctuations,
- Reynolds stresses,
- Flow direction and flow angle (in the case of two- or three-dimensional probes),
- Flow temperature, and
- Reversed flow detection.

The operating principle of the hot-wire anemometer (see Fig. 4.6) is based on convective heat transfer from a fine wire (typically a few microns in diameter) made of tungsten, platinum, or platinum-based alloys. The wire is mounted between two prongs and placed directly in the fluid stream. Any change in the flow characteristics alters the rate of heat dissipation from the wire, which is then detected and translated into velocity information.

Notably, while the sensitivity of other measurement techniques tends to diminish at low velocities, the sensitivity of hot-wire anemometry increases in such conditions. This makes it especially suitable for the study of low-speed flow regimes.

As previously emphasized, the hot-wire anemometer is one of the principal tools for measuring instantaneous flow velocity. Due to its unique advantages, its primary

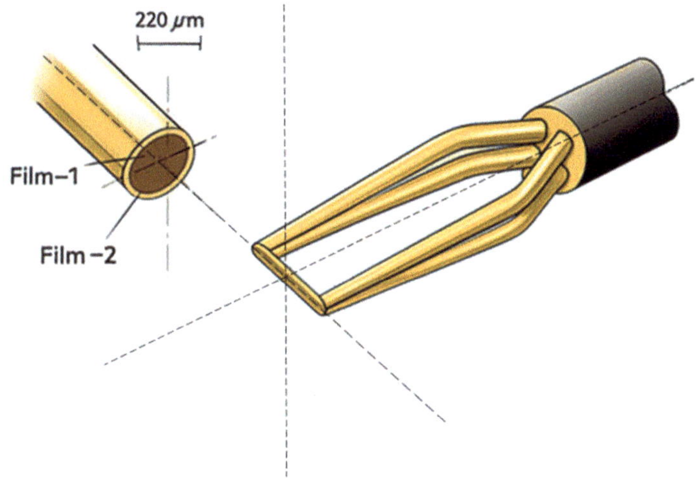

Fig. 4.6 Schematic view of a split-film sensor

application lies in experimental investigations of complex flow behaviors across diverse scenarios. A key example is the study of airflow around building models and the accurate determination of boundary layer characteristics.

The boundary layer, which forms due to the relative motion between a solid body and a fluid, is a fundamental concept in fluid mechanics that has been the subject of extensive past and ongoing research. This importance stems from the boundary layer's direct and indirect influence on a wide range of fluid dynamic phenomena, such as:

- Flow separation,
- Wake formation,
- Skin friction, and
- Lift and drag forces.

By accurately characterizing and controlling boundary layer development, the design and optimization of physical systems—such as aircraft, automobiles, HVAC systems, and heat exchangers—can be greatly improved.

The boundary layer forms as a result of the relative motion between a solid body and a fluid (or between two different fluids such as air and water). If the solid surface is stationary, the velocity at the wall is zero due to the no-slip condition, and it gradually increases with distance from the wall, asymptotically approaching the free-stream velocity U_∞. The boundary layer thickness may vary from a few millimeters to several tens of meters depending on the flow conditions. In the experimental investigations presented in this study, the boundary layer thickness has been measured and analyzed.

The boundary layer can generally be classified into two types: laminar and turbulent. A laminar boundary layer is characterized by a stable, two-dimensional flow structure, while a turbulent boundary layer exhibits unstable, three-dimensional flow behavior. The region in which the transition from laminar to turbulent flow occurs is referred to as the transitional boundary layer. In most natural and engineering applications, the boundary layer is predominantly turbulent.

Boundary layer investigations can be conducted using both numerical simulations and experimental techniques. Among the experimental methods, the hot-wire anemometer is particularly effective for profiling velocity fields within boundary layers. This technique has also been employed in the present study.

The boundary layer thickness is conventionally defined as the distance from the wall at which the streamwise velocity component u reaches 99% of the free-stream velocity U_∞. That is, the edge of the boundary layer is located at the position where:

$$u = 0.99\, U_\infty \tag{4.5}$$

To experimentally determine the boundary layer thickness, velocity sensors must be positioned perpendicular to the wall and measurements should be taken starting from close to the surface. The vertical location at which the above velocity criterion is met corresponds to the boundary layer thickness.

In low- and moderately turbulent flows (with turbulence intensity below 25%), hot-wire anemometry provides significantly higher measurement accuracy than alternative techniques. Several advantages of hot-wire anemometers compared to other velocity measurement tools are summarized below:

- Ease of use: Calibration, data acquisition, and post-processing are straightforward.
- Superior low-velocity performance: While the accuracy of many velocity sensors diminishes at low speeds, the precision of hot-wire anemometers improves as velocity decreases.
- High accuracy: Under optimal conditions, the measurement uncertainty of the hot-wire anemometer can be as low as 0.1–0.2% of the measured value.

The measurement capabilities of the hot-wire anemometer extend across a wide range of flow velocities, making it suitable for both low-speed and moderate-to-high-speed applications. The frequency response of a typical hot-wire system is approximately 50 kHz, though with appropriate electronic circuitry, this can be extended to several hundred kilohertz. The system also exhibits a low signal-to-noise ratio, ensuring stable and accurate voltage output. The measured data are provided in the form of a continuous analog voltage signal, which simplifies data acquisition and subsequent signal processing.

Despite these advantages, proper handling and operational care are essential for reliable use of hot-wire probes. The following considerations must be observed:

- Flow contamination (e.g., dust or oil particles) can alter the sensor's calibration characteristics by affecting heat transfer, resulting in an increased response time. In such cases, the sensor must be carefully cleaned and recalibrated before further use.
- The hot-wire sensor itself is extremely delicate and susceptible to damage. However, with careful handling, it can have a long operational lifespan. The probe

consists of two parallel nickel-coated wires mounted on a quartz fiber substrate, with a typical wire diameter of 200 μm and a length of 3 mm. The supporting shaft is made of ceramic, with a diameter of 2.3 mm, and includes gold-coated connector pins that interface with the probe base using a dual-prong plug configuration.
- In high-speed flows, fine particulate matter can strike and break the sensing wire. This risk can be mitigated by incorporating filters into the upstream flow path.
- Overheating the sensor beyond its rated temperature can cause permanent damage. Therefore, proper thermal regulation is critical.
- The probe must be mounted using a specially designed holder, which should be carefully positioned to minimize flow disturbances. Poor installation can generate unwanted turbulence that affects measurement accuracy.
- During calibration, the sensor is aligned perpendicular to the flow direction. Maintaining this orientation during actual measurements ensures consistency and improves the accuracy and repeatability of the results.

4.4 Calibration of the Hot-Wire Anemometer

As previously discussed, the primary function of the hot-wire anemometer is to measure the instantaneous velocity of a fluid flow, from which flow quantities such as mean velocity, turbulence intensity, and Reynolds stresses can be derived. The output signal of the anemometer is a voltage that exhibits a nonlinear relationship with the flow velocity. Given the high sensitivity and precision of the hot-wire system, even minor changes in flow conditions, probe orientation, sensor contamination, or probe displacement can significantly affect the output voltage. Therefore, accurate calibration is essential prior to any experimental measurements.

The objective of calibration is to establish a precise correlation between flow velocity and the corresponding output voltage of the sensor. This process requires a known and controlled velocity field, so that the sensor's voltage output can be recorded at discrete flow speeds. To generate these reference velocities, a calibration apparatus is required—typically either a wind tunnel or a compressed air chamber equipped with a nozzle. In the present study, calibration was performed using a laboratory wind tunnel.

To calibrate the hot-wire anemometer, the target velocity range is first determined and divided into at least ten discrete intervals. At each interval, the output voltage is recorded. A critical requirement during calibration is that the configuration of the probe during calibration must exactly match its configuration during actual measurement. Due to the sensitivity of the system, it is strongly recommended that the same wind tunnel be used both for calibration and for the subsequent experiments. In other words, calibration and measurement should ideally be performed under identical environmental and geometric conditions to eliminate systematic errors.

The calibration procedure employed in this research is described as follows: The hot-wire anemometer measures the instantaneous velocity at the specific spatial location of the probe. To capture velocity distributions or spatial velocity gradients, a precise traversing mechanism is required to reposition the probe along defined paths. Furthermore, for multi-component hot-wire probes (e.g., two- or three-wire configurations), a rotational stage is needed to orient the probe with respect to the local flow direction. This rotational mechanism also facilitates the calibration technique based on the vortex shedding behind a cylindrical rod, which is widely used in aerodynamic studies. In this technique, the probe is accurately positioned in the near-wake region downstream of a bluff body, where vortex streets form. The frequency and intensity of the shed vortices provide a reliable calibration reference, particularly for multi-component flow measurements.

The probe traversing system is actuated using stepper motors, which provide precise angular motion typically with a resolution of approximately 1° per step. When the mechanical components of the traversing mechanism are fabricated with sufficient precision, the probe can be repositioned with high accuracy at the desired measurement locations. To ensure such accuracy, all lead screws and guide shafts must be manufactured with minimal backlash or clearance. Moreover, smaller thread pitches in the lead screws enhance the positional accuracy and resolution of the probe movement.

Stepper motors are controlled via electronic drivers, which receive their motion commands from a dedicated control software interfaced with a microcontroller unit. This setup allows for programmable, repeatable motion control suitable for automated data acquisition in wind tunnel experiments.

Designing and installing a probe traversing mechanism is relatively straightforward for wind tunnels with a test section cross-sectional area larger than 1.5 m^2, as the mechanism can be placed directly inside the test section without significantly disturbing the flow. However, in small-scale wind tunnels—such as the one used in the present study—internal placement of the probe mechanism is not feasible. This is primarily due to the increased blockage ratio, which refers to the ratio of the cross-sectional area occupied by the model and instrumentation to the total test section area. Excessive blockage distorts the flow field and introduces measurement errors. Therefore, in such cases, the probe positioning system must be mounted externally, outside the test section. Figure 4.7 illustrates the external probe traversing system as viewed from above the wind tunnel.

As previously discussed, the goal of calibration is to establish an accurate correlation between flow velocity and the voltage output of the hot-wire anemometer. Several calibration techniques exist, which can be categorized based on the velocity range of interest. This categorization is crucial because varying the flow velocity alters the flow regime and, consequently, the dimensionless parameters influencing the sensor response, such as the Reynolds number and Mach number. A proper calibration method must account for these changes to ensure the accuracy and reliability of the velocity measurements.

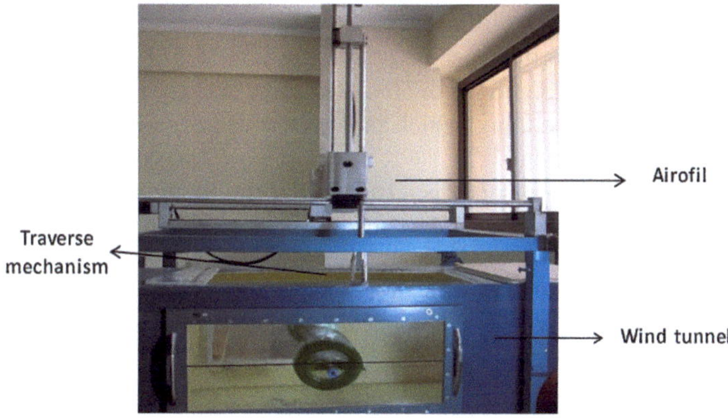

Fig. 4.7. Probe transfer mechanism outside the test section in the open-circuit wind tunnel at Shiraz University, Iran

4.4.1 Calibration Using Regular Vortex Shedding

This calibration technique is based on the generation of periodic vortex shedding downstream of a slender circular cylinder—commonly referred to as a Kármán vortex street—placed in a uniform flow. Figure 4.8 illustrates the flow regimes around a cylinder for various Reynolds numbers, defined as $\mathrm{Re} = \frac{Ud}{\nu}$, where d is the cylinder diameter, U is the free-stream velocity, and ν is the kinematic viscosity of the fluid.

At very low Reynolds numbers, the flow remains symmetric about the cylinder with no wake formation. As Re increases beyond approximately 30–40, boundary layers begin to form on the surface, and a steady laminar wake appears. When Re reaches the range of 40–50, vortex shedding initiates, manifesting as an oscillatory

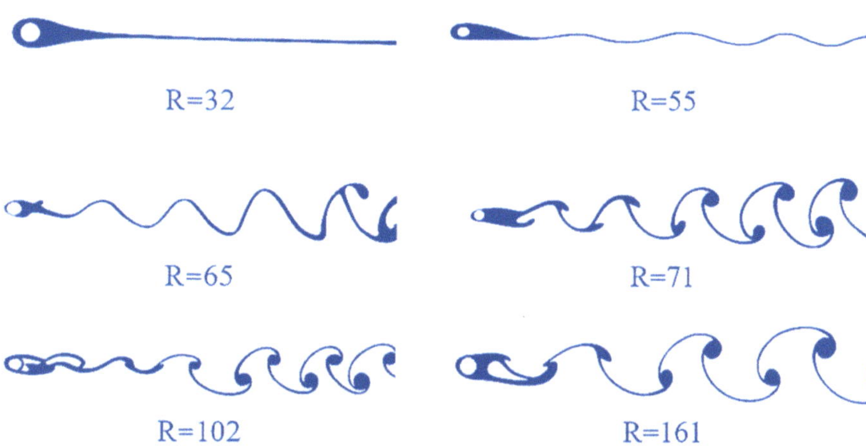

Fig. 4.8 Flow around a circular cylinder at different Reynolds numbers

flow downstream. At Re $\approx$ 140, the vortices become clearly distinguishable and grow in size. Periodic vortices persist up to approximately Re $\approx$ 2000. Therefore, within the Reynolds number range of 50–2000, the wake downstream of the cylinder exhibits regular vortex shedding, which can be recorded as sinusoidal voltage signals using a hot-wire anemometer.

The vortex shedding frequency is a function of the Reynolds number and can be approximated by the following piecewise empirical correlations:

$$F = 0.212\,\text{Re} - 4.5 \quad \text{for } 50 < \text{Re} < 150 \tag{4.6}$$

$$F = 0.212\,\text{Re} - 2.7 \quad \text{for } 300 < \text{Re} < 2000 \tag{4.7}$$

Here, F is the dimensionless frequency, defined as:

$$F = \frac{fd^2}{\nu} \tag{4.8}$$

where f is the vortex shedding frequency in Hz, d is the cylinder diameter, and ν is the kinematic viscosity of the fluid.

Alternatively, the vortex shedding frequency can be expressed using the Strouhal number:

$$\text{St} = \frac{fd}{U} \tag{4.9}$$

With the following empirical relations:

$$\text{St} = 0.212\left(1 - \frac{21.2}{\text{Re}}\right) \quad \text{for } 50 < \text{Re} < 150 \tag{4.10}$$

$$\text{St} = 0.212\left(1 - \frac{12.7}{\text{Re}}\right) \quad \text{for } 300 < \text{Re} < 2000 \tag{4.11}$$

Using a hot-wire anemometer, the vortex shedding frequency downstream of the cylinder can be experimentally measured. With the known values of d and ν, the dimensionless frequency F can be computed. By inverting Eq. (4.1) or using Eq. (4.2), the Reynolds number—and subsequently the flow velocity U—can be determined. This method provides an indirect calibration of the hot-wire sensor by correlating its output to a known periodic flow structure.

To extract the vortex frequency f, either a spectrum analyzer or a data acquisition system must be used. In the latter case, the voltage signal from the hot-wire is sampled and transferred to a computer, where a Fast Fourier Transform (FFT) is applied to convert the time-domain signal into its frequency-domain representation. The dominant frequency corresponds to the vortex shedding.

Since vortex shedding frequency depends on both the cylinder diameter and the free-stream velocity, the first step in this calibration method is to define the desired velocity range $U_{\min}$ to $U_{\max}$. An appropriate cylinder diameter must then be selected using Eq. (4.1) or (4.2). Decreasing the cylinder diameter increases the frequency of vortex shedding for a given velocity range.

For example, to calibrate for a minimum velocity of 0.5 m/s, assuming a fluid viscosity $\nu = 1.5 \times 10^{-5}\,\text{m}^2/\text{s}$, the cylinder diameter should be at least 1.5 mm, yielding a vortex frequency of approximately 67 Hz. This setup allows accurate calibration for flow velocities up to 20 m/s. Alternatively, using a 0.6 mm diameter cylinder, the method can be extended to calibrate for velocities up to 50 m/s, resulting in vortex frequencies on the order of 17 kHz. To capture such high-frequency signals accurately, the sampling frequency must comply with the Nyquist criterion [6].

To ensure clear and undistorted vortex signals, the hot-wire probe must be positioned at an appropriate downstream distance from the cylinder. Figure 4.9 shows sample hot-wire voltage signals recorded at two locations behind the cylinder. Improper positioning of the probe can lead to signal distortion and poor sinusoidal resolution of the vortex pattern.

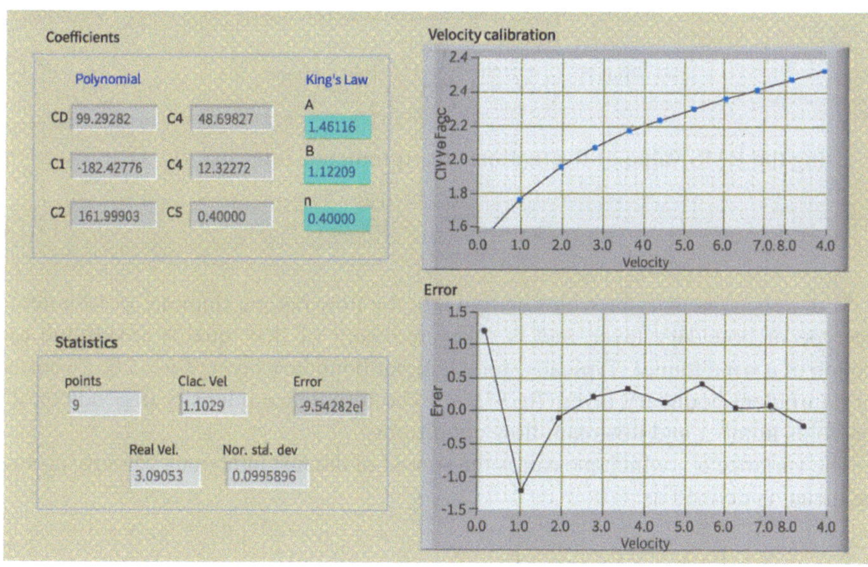

Fig. 4.9. A sample calibration curve of the hot-wire anemometer

4.5 Analysis of Measured Instantaneous Velocity

Since the hot-wire anemometer measures the instantaneous velocity of the flow, it is necessary to statistically analyze the recorded data to extract meaningful quantities. One of the most fundamental statistical parameters is the mean flow velocity, defined by:

$$U_{\text{mean}} = \frac{1}{N} \sum_{n=1}^{N} U(n) \tag{4.12}$$

where $U(n)$ is the instantaneous velocity at sample n, N is the total number of samples, and U_{mean} is the mean velocity.

By relocating the probe across the flow field using the probe translation mechanism, the spatial distribution of velocity—such as boundary layer profiles, wake regions, and shear layers—can be determined.

The root-mean-square (RMS) of velocity fluctuations, which indicates the intensity of turbulence, is calculated as:

$$u_{\text{rms}} = \sqrt{\frac{1}{N-1} \sum_{n=1}^{N} (U(n) - U_{\text{mean}})^2} \tag{4.13}$$

Using the RMS value of fluctuations, the turbulence intensity Tu is expressed as:

$$Tu = \frac{u_{\text{rms}}}{U_{\text{mean}}} \times 100\% \tag{4.14}$$

Turbulence intensity provides insight into the flow regime (laminar or turbulent), the type of boundary layer, and is a key indicator of flow quality inside the test section of a wind tunnel. Typically, for a laminar boundary layer, $Tu < 2\%$, whereas for a turbulent boundary layer, $Tu > 5\%$. The turbulence intensity depends on the Reynolds number and upstream flow conditions.

The variance σ^2, which measures the spread of the instantaneous velocity around the mean, is defined as:

$$\sigma^2 = \frac{1}{N-1} \sum_{n=1}^{N} (U(n) - U_{\text{mean}})^2 \tag{4.15}$$

In addition to the mean and variance, two higher-order central moments are often used for flow characterization:

- Skewness S, which represents the third central moment, is a measure of asymmetry in the velocity distribution:

$$S = \frac{1}{N\sigma^3}\sum_{n=1}^{N}(U(n) - U_{\text{mean}})^3 \quad (4.16)$$

- Kurtosis K, the fourth central moment, indicates the peakedness or flatness of the velocity probability distribution:

$$K = \frac{1}{N\sigma^4}\sum_{n=1}^{N}(U(n) - U_{\text{mean}})^4 \quad (4.17)$$

Skewness quantifies how symmetric the velocity fluctuations are around the mean velocity, while kurtosis provides a measure of how concentrated the fluctuations are—i.e., how sharp or flat the distribution is compared to a Gaussian.

The instantaneous velocity signal can be analyzed both in the time domain and in the frequency domain. Frequency-domain analysis reveals how the turbulent energy is distributed across different frequencies, which is essential for understanding the energy spectrum of the turbulent flow.

4.6 Effect of Turbulence Intensity on Flow Velocity Measurements

In most practical situations, turbulence in fluid flow is inherently three-dimensional. When the turbence intensity is low, the velocity components perpendicular to the main flow direction can be neglected, and for a one-dimensional flow field, the transverse velocity components can be assumed zero, i.e., $V = W = 0$. Under such conditions, the instantaneous velocity is represented as:

$$U(n) = \overline{U} + u(n) \quad (4.18)$$

However, in realistic one-dimensional flows—such as turbulent boundary layers or wake regions—velocity fluctuations are not negligible and may introduce errors in velocity measurements. Even if the mean transverse components $\overline{V} = \overline{W} = 0$, the fluctuating components v and w still exist and influence the heat transfer characteristics.

Since the directional components of the flow field contribute non-uniformly to heat transfer and sensor output, the effective velocity measured by the hot-wire anemometer is expressed as:

$$V_e^2 = U_N^2 + k^2U_T^2 + h^2U_B^2 \quad (4.19)$$

Here:

- $U_N, U_T and U_B$ are the velocity components in the normal, transverse (yaw), and binormal (pitch) directions,
- k and h are the sensitivity coefficients corresponding to the yaw angle α and pitch angle β of the flow relative to the probe orientation (see corresponding schematic).

Given that $h^2 \ll k^2$, the contribution of fluctuations in the transverse direction (associated with V) to the measurement of the streamwise velocity U can be considered negligible.

For one-dimensional flows with $\overline{V} = \overline{W} = 0$, and assuming that the instantaneous velocity can be decomposed as $U(t) = \overline{U} + u(t)$, Eq. (4.8) simplifies to:

$$V_e^2 = U^2 + k^2v^2 + h^2w^2 \tag{4.20}$$

Taking the square root yields:

$$V_e = \sqrt{(U+u)^2 + k^2v^2 + h^2w^2} \tag{4.21}$$

Expanding this expression as a binomial series gives:

$$V_e = (U+u) + \frac{1}{2}\frac{k^2v^2 + h^2w^2}{U+u} + \frac{1}{8}\frac{(k^2v^2 + h^2w^2)^2}{(U+u)^3} + \cdots \tag{4.22}$$

Assuming $h^2 \ll k^2$, the expression simplifies to:

$$V_e = \overline{U}\left[1 + \frac{u}{\overline{U}} + \frac{1}{2}\frac{h^2w^2}{\overline{U}(\overline{U}+u)} + \cdots\right] \tag{4.23}$$

If the fluctuation $u \ll \overline{U}$, a further simplified expression for the measured mean velocity becomes:

$$\overline{U}_m = \overline{U}\left(1 + \frac{1}{2}\frac{h^2\overline{w^2}}{\overline{U}^2}\right) \tag{4.24}$$

Likewise, the measured velocity fluctuation intensity is derived from:

$$\overline{u_m^2} = \overline{u^2}\left(1 + h^2 R_{uw^2} Tu\right) \tag{4.25}$$

where

$$Tu = \frac{\sqrt{\overline{u^2}}}{\overline{U}} \tag{4.26a}$$

$$R_{uw^2} = \frac{\overline{uw^2}}{(\overline{u^2})^{3/2}} \tag{4.26b}$$

Table 4.1 illustrates the effect of turbulence intensity on the measured mean velocity

$\overline{U}_m/\overline{U}$	Turbulence intensity $Tu(\%)$
1.006	10
1.02	20
1.05	30

In these equations,

- $\overline{U}_m$ and $\overline{u_m^2}$ are the measured mean velocity and measured turbulence intensity, respectively,
- $\overline{U}$ and $\overline{u^2}$ are the actual mean velocity and true turbulence intensity, respectively.

Equations (4.25) and (4.26a, 4.26b) demonstrate that when turbulence intensity is high, the measured values deviate from the actual values, introducing significant error into the measurements.

Assuming homogeneous turbulence such that $\overline{u^2} = \overline{v^2} = \overline{w^2}$, the error induced by turbulence on the mean velocity measurement is represented in Table 4.1. As shown, the deviation is negligible at low turbulence levels, but for $Tu = 30\%$, the measured mean velocity is overestimated by approximately 5%.

4.7 Estimation of Experimental Error (Measurement Uncertainty)

The occurrence of human and equipment-related errors in experimental procedures is virtually unavoidable. If these errors are not identified and properly quantified, the reliability of the experimental results becomes questionable. Among the most common sources of error encountered in laboratory measurements are visual misreading, sensor inaccuracies, and computational errors. However, given that multiple factors influence any measurement process, determining the exact value of the overall uncertainty is often challenging, and in many cases, practically impossible.

Despite these limitations, approximate estimates of measurement uncertainty can be obtained by applying established analytical methods that account for key influencing variables. In the present study, the only directly measured parameter is the mean flow velocity, and its associated uncertainty is evaluated using the method proposed by Adams.

According to this method, if a physical quantity P is a function of independent variables q, r, and s, the total uncertainty in P resulting from small variations in these parameters can be approximated using the following general expression:

$$(\delta P)^2 = \left|\frac{\partial P}{\partial q}\right|^2 (\delta q)^2 + \left|\frac{\partial P}{\partial r}\right|^2 (\delta r)^2 + \left|\frac{\partial P}{\partial s}\right|^2 (\delta s)^2 \tag{4.27}$$

The corresponding relative (percentage) error is given by:

$$\% \text{ ERROR} = \frac{\delta P}{P} \times 100 \tag{4.28}$$

In this experiment, the variables that influence the measured flow velocity are the voltage output of the hot-wire anemometer and the ambient temperature. The ambient temperature is recorded using a laboratory-grade thermometer, while the voltage is obtained from the calibration curve of the hot-wire probe. The precision of voltage measurements depends on the resolution of the data acquisition (DAQ) system used.

When calibrating the hot-wire sensor, methods such as regular vortex shedding may be employed. In such cases, the free-stream velocity U_∞, which is used as a calibration reference, also contains measurement uncertainty. This velocity is typically estimated from the empirical correlation between the Strouhal number S_N and the Reynolds number Re, which is expressed as:

$$S_N = 0.212\left(1 - \frac{21.2}{Re}\right), \quad 50 < Re < 150 \tag{4.29}$$

$$S_N = 0.212\left(1 - \frac{12.7}{Re}\right), \quad 300 < Re < 2000 \tag{4.30}$$

The Strouhal number can also be defined by the non-dimensional frequency relationship:

$$S_N = \frac{fd}{U_\infty} \tag{4.31}$$

Hence, the uncertainty in S_N, as a result of uncertainties in the frequency f, body diameter d, and velocity U_∞, can be computed using:

$$\delta S_N = \sqrt{\left(\frac{d}{U_\infty}\delta f\right)^2 + \left(\frac{f}{U_\infty}\delta d\right)^2 + \left(\frac{-fd}{U_\infty^2}\delta U_\infty\right)^2} \tag{4.32}$$

Since both the Strouhal and Reynolds numbers are functions of the flow velocity, this leads to an implicit relationship between the uncertainty and the flow velocity itself. To resolve this, the following alternative expressions based on dimensionless frequency F are used:

$$F = 0.212Re^{-4.5}, \quad 50 < Re < 150 \tag{4.33}$$

$$F = 0.212Re^{-2.7}, \quad 300 < Re < 2000 \tag{4.34}$$

$$F = \frac{fd^2}{\nu} \tag{4.35}$$

Using these relations, the uncertainty in F can be explicitly propagated to determine the uncertainty in U_∞, as they are directly related via kinematic viscosity ν, vortex shedding frequency f, and body diameter d.

To estimate the uncertainty in the measured mean velocity U, the voltage–velocity–temperature calibration relationship derived from King's Law is applied. The complete set of equations is as follows:

$$U = f(E, T_\infty) \Rightarrow \delta U = \left|\frac{\partial U}{\partial E}\right| \delta E + \left|\frac{\partial U}{\partial T_\infty}\right| \delta T_\infty \tag{4.36}$$

$$E^2 = \left(A + BU^n\right)(T_m - T_\infty) \tag{4.37}$$

$$U = \left[\frac{1}{B}\left(\frac{E^2}{T_m - T_\infty} - A\right)\right]^{1/n} \tag{4.38}$$

where

- E is the measured voltage [V],
- T_∞ is the ambient temperature [°C] (typically 27 °C)
- T_m is the wire temperature of the hot-wire probe [°C] (typically 250 °C),
- U is the mean flow velocity [m/s], and
- $A, B,$ and n are the empirical calibration constants, determined via curve fitting.

These equations allow the computation of both the velocity and its associated uncertainty, using known values of voltage and temperature, and the calibration constants obtained for each sensor.

4.8 Two-Dimensional Velocity Measurement Using Hot-Wire Anemometry

Assuming the vertical velocity component is negligible (i.e., $W = 0$), the flow velocity vector comprises two components, U (streamwise) and V (cross-stream). The effective (measured) velocity V_e is given by:

$$V_e^2 = U^2 + K^2 V^2 \tag{4.39}$$

where K is the sensor's directional sensitivity coefficient with respect to the yaw angle α, determined through directional calibration.

To determine the components U and V, it is necessary to obtain two independent measurements of effective velocity, denoted as V_{e1} and V_{e2}. These are obtained by rotating a single hot-wire probe (e.g., SN or SY type) to two different yaw angles, $+\overline{\alpha}$ and $-\overline{\alpha}$, and recording the corresponding output voltages E_1 and E_2. These voltages are then converted into effective velocities using the probe's calibration curve.

Alternatively, a cross-wire (X-type) probe containing two hot wires can simultaneously measure both components U and V by capturing the voltages E_1 and E_2 from each wire and converting them to velocities via calibration.

Measurement Using a Single-Wire Probe:

The measurement procedure for two-dimensional flow using a single-wire probe is as follows:

1. Mount the probe at an initial yaw angle $\overline{\alpha}$ relative to the principal coordinate axes.
2. Measure the instantaneous voltage E_1, correct it for amplification and offset, apply the temperature correction, and convert it into effective velocity V_{e1}.
3. Rotate the probe to yaw angle $-\overline{\alpha}$ and repeat the measurement to obtain V_{e2}.

For a probe at yaw angle $\overline{\alpha}$, the velocity components relative to the sensor orientation are:

$$U_T = U\sin\overline{\alpha} + V\cos\overline{\alpha} \tag{4.40}$$

$$U_N = U\cos\overline{\alpha} - V\sin\overline{\alpha} \tag{4.41}$$

where U_T and U_N are the tangential and normal components, respectively.

Substituting into the effective velocity formula:

$$V_{e1}^2 = (U\cos\overline{\alpha} - V\sin\overline{\alpha})^2 + k^2(U\sin\overline{\alpha} + V\cos\overline{\alpha})^2 \tag{4.42}$$

$$V_{e2}^2 = (U\cos\overline{\alpha} + V\sin\overline{\alpha})^2 + k^2(U\sin\overline{\alpha} - V\cos\overline{\alpha})^2 \tag{4.43}$$

Expanding and rearranging the terms yields:

$$V_{e1}^2 = A_1U^2 + A_2V^2 - A_3UV \tag{4.44}$$

$$V_{e2}^2 = A_1U^2 + A_2V^2 + A_3UV \tag{4.45}$$

where

$$A_1 = \cos^2\overline{\alpha} + k^2\sin^2\overline{\alpha} \tag{4.46}$$

$$A_2 = \sin^2\overline{\alpha} + k^2\cos^2\overline{\alpha} \tag{4.47}$$

$$A_3 = \left(1 - k^2\right)\sin2\overline{\alpha} \tag{4.48}$$

Solving this system of equations allows one to compute the components U and V.

Simplified Expressions under Low Turbulence Intensity:

Assuming small turbulence intensity, the effective velocities can be approximated as:

$$V_{e1} = f(\overline{\alpha})\left[U - g(\overline{\alpha})V\right] \tag{4.49}$$

$$V_{e2} = f(\overline{\alpha})\left[U + g(\overline{\alpha})V\right] \tag{4.50}$$

where

$$f(\overline{\alpha}) = \sqrt{\cos^2\overline{\alpha} + k^2\sin^2\overline{\alpha}} \tag{4.51}$$

$$g(\overline{\alpha}) = \frac{\left(1 - k^2\right)\cos^2\overline{\alpha}}{\cos^2\overline{\alpha} + k^2\sin^2\overline{\alpha}}\tan\overline{\alpha} \tag{4.52}$$

Statistical Analysis Using Triple-Angle Measurements:

Using voltage readings at three probe orientations—$\alpha = 0°$, $+\overline{\alpha}$, and $-\overline{\alpha}$—one can estimate turbulence quantities:

- At $\alpha = 0°$: $\overline{V}_e = \overline{U}$ and $\overline{V}_e^2 = \overline{u^2}$
- At $\alpha = \pm\overline{\alpha}$:

$$\overline{V}_{e1}^2 = f^2(\overline{\alpha})\left[\overline{u^2} + g^2(\overline{\alpha})\overline{v^2} - 2g(\overline{\alpha})\overline{uv}\right] \tag{4.53}$$

$$\overline{V}_{e2}^2 = f^2(\overline{\alpha})\left[\overline{u^2} + g^2(\overline{\alpha})\overline{v^2} + 2g(\overline{\alpha})\overline{uv}\right] \tag{4.54}$$

Solving these equations yields the mean velocities $\overline{U}$, $\overline{V}$, as well as turbulence quantities $\overline{u^2}$, $\overline{v^2}$, and the Reynolds shear stress $\overline{uv}$.

4.8.1 Uncertainty Estimation in Two-Dimensional Measurements

To assess measurement reliability, the longitudinal and transverse components of the velocity field over a flat plate are compared with validated datasets. The deviation between experimental and reference values should remain below 15%. Deviations above this threshold are generally attributed to:

- Differences in Reynolds number,
- Variations in surface roughness (wall roughness height), and
- Differences in wind tunnel configuration or flow quality.

4.9 Construction of a Scaled Model for Wind Tunnel Testing

To conduct reliable laboratory experiments, a geometrically accurate and scaled-down physical model of the target building and its surrounding urban block must be fabricated according to the predefined scaling parameters discussed earlier. It is crucial that the model replicates the original geometry with high fidelity and is constructed from materials that closely mimic the aerodynamic and thermal behavior of the actual construction materials [9, 10].

Figure 4.10 illustrates a scaled model of an urban neighborhood tested inside a wind tunnel facility. Similarly, Fig. 4.11 shows an isolated building model positioned in the test section of the wind tunnel. It is important to note that, for accurate evaluation of wind effects on a particular building, the influence of neighboring structures—which significantly affect the flow characteristics—must be accounted for in the model. Neglecting surrounding buildings can lead to an unrealistic representation of the wind environment and flow separation patterns.

Moreover, full-scale wind tunnel facilities do exist globally, where life-size structures can be experimentally tested under controlled wind conditions. However, for most architectural and urban flow studies, especially when analyzing wind behavior around buildings and neighborhoods, scaled-down physical models are typically sufficient and widely accepted.

4.10 Fabrication of Vorticity Generator and Surface Roughness Elements

Accurate simulation of wind flow over buildings or urban neighborhoods in wind tunnel experiments requires two essential components: (1) a precisely scaled physical model, as previously discussed, and (2) realistic reproduction of the atmospheric wind profile. The latter ensures that the approaching wind flow matches the actual velocity and turbulence profiles found in the atmospheric boundary layer.

To replicate the atmospheric boundary layer (ABL) within the wind tunnel, various techniques are available. Among the most effective is the method introduced by Counihan, which involves the use of a vorticity generator to induce shear and turbulence at the inlet section of the test chamber (see in Figs. 4.5d and 4.12).

Accordingly, a custom-designed vorticity generator should be fabricated to match the cross-sectional dimensions of the wind tunnel. For studies involving urban

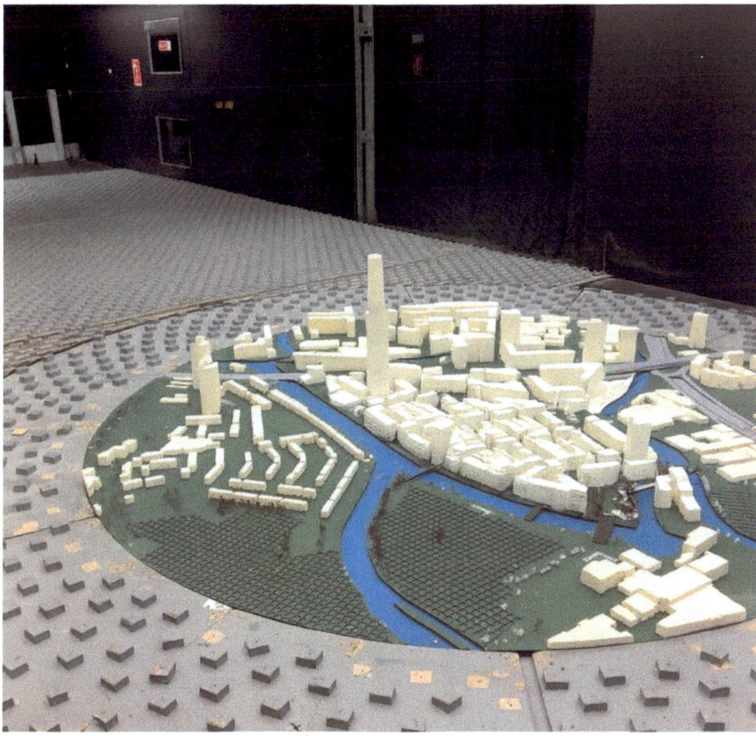

Fig. 4.10 A sample urban context in a wind tunnel

models and building aerodynamics, a generator with 4–5 vanes is typically sufficient (Fig. 4.13a,b). It is recommended that the base structure be constructed from heavy steel for stability, while the vanes themselves may be fabricated from plexiglass sheets due to their low inertia and ease of adjustment.

In addition to the vorticity generator, surface roughness elements must also be installed on the tunnel floor to emulate the roughness of natural terrain or upstream obstacles (see in Figs. 4.13c and 4.14). These roughness elements are typically metallic rectangular prisms of varying dimensions, positioned between the vorticity generator and the test model. This configuration facilitates realistic boundary layer development and turbulence intensity appropriate for urban flow simulations.

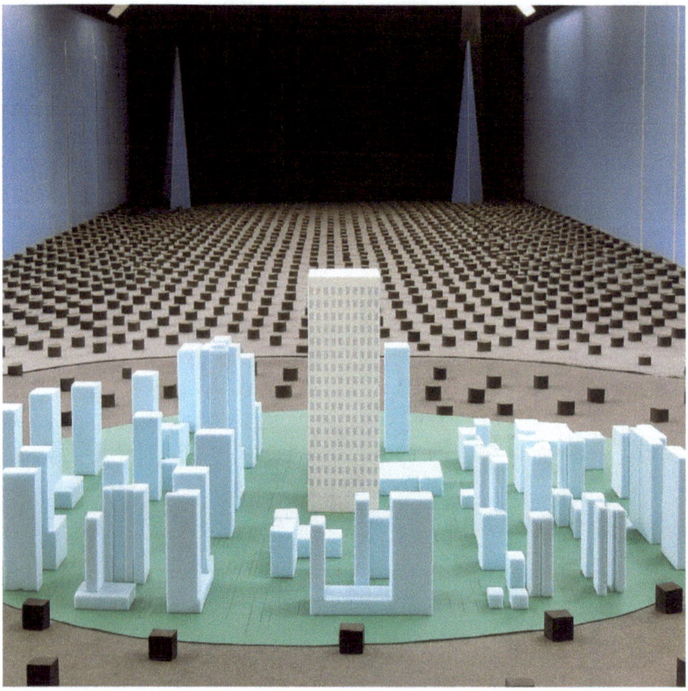

Fig. 4.11 Building model with surrounding structures in the wind tunnel

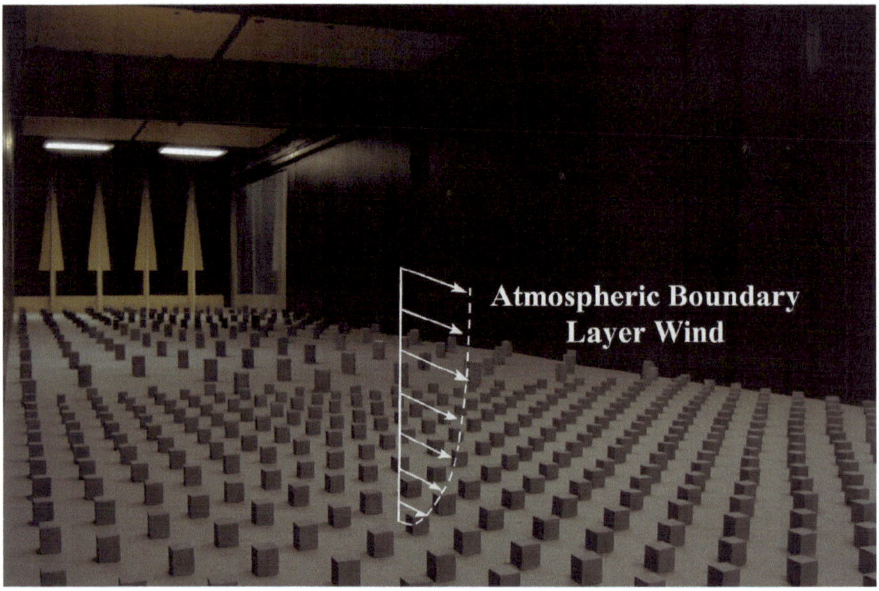

Fig. 4.12 Vorticity and surface roughness in an open-circuit wind tunnel for boundary layer generation

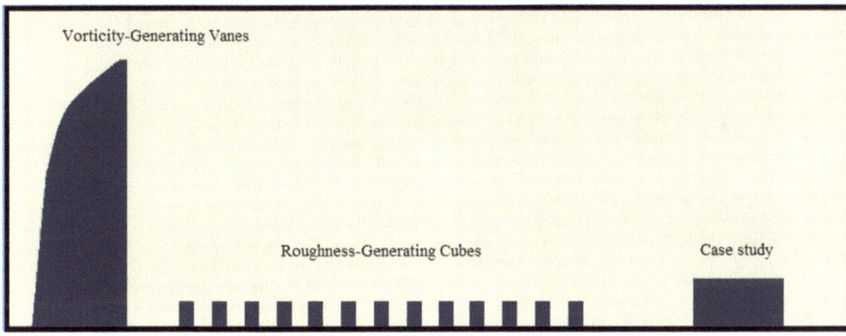

a) Cross-sectional view of the wind tunnel at the test section showing roughness blocks and vorticity-generating vanes

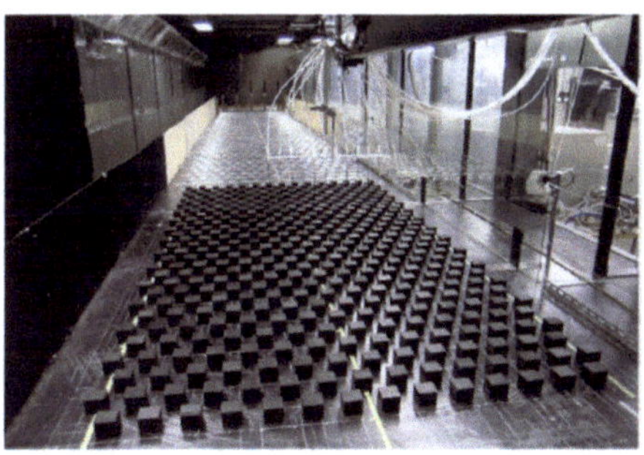

c) Roughness-Generating Cubes

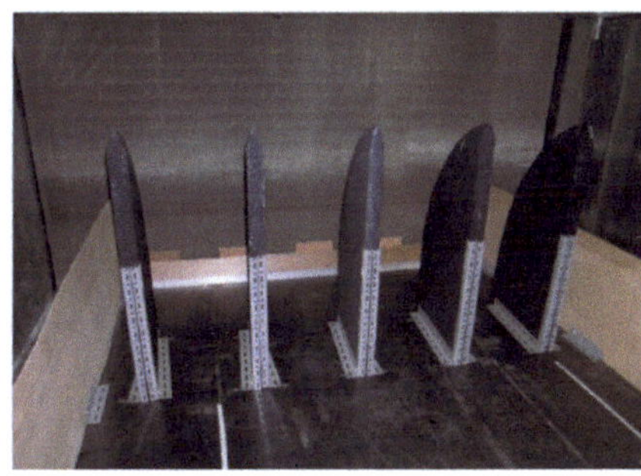

b) Vorticity-Generating Vanes

Fig. 4.13. Cross-section and components for wind profile generation in an open-circuit wind tunnel. **a** Cross-sectional view of the wind tunnel at the test section showing roughness blocks and vorticity-generating vanes. **b** Vorticity-Generating Vanes. **c** Roughness-Generating Cubes

4.11 Experimental Data Output

4.11.1 Numerical Data Acquisition

Once the scaled model is positioned inside the wind tunnel test section, the incident wind profile can be directed toward the model, and experimental data can be extracted. Using appropriate velocity probes and turbulence sensors, quantitative data such as instantaneous and time-averaged velocity components, turbulence intensities, and Reynolds stresses can be recorded.

These experimental results are frequently compared with CFD simulation outputs for validation and calibration purposes. For example, identical boundary conditions and geometric configurations are replicated in a CFD platform such as ANSYS Fluent or COMSOL, and the simulation results are directly compared with the wind tunnel measurements. This comparative analysis is essential for validating the numerical model or further investigating flow characteristics.

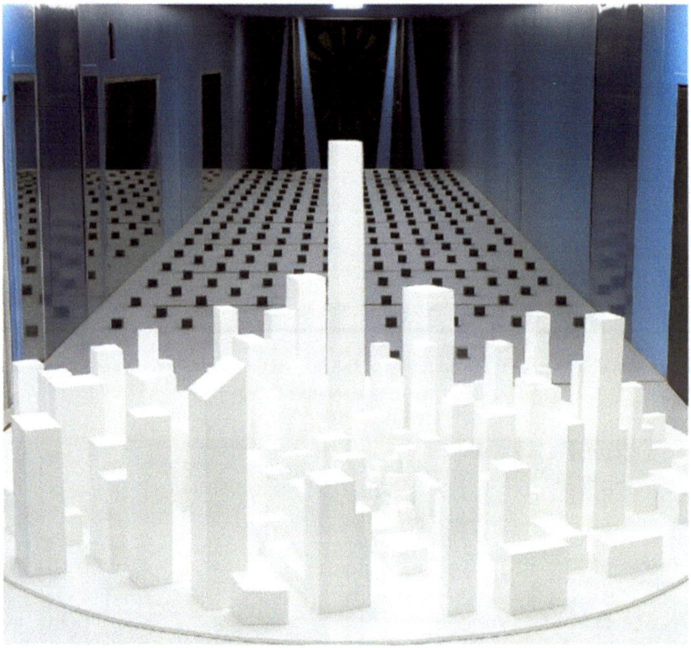

Fig. 4.14 Experimental setup within the test section including surface roughness elements and vorticity-generating vanes

Figure 4.5 demonstrates this validation process, where an architectural room equipped with a building was tested both experimentally and numerically, and the corresponding flow fields were compared for consistency (Fig. 4.5b).

4.11.2 Flow Visualization Using Smoke Generator

In addition to numerical measurements, qualitative flow visualization techniques are employed to investigate flow patterns and turbulent structures over the model. A smoke generator is used to inject visible tracer particles into the incoming flow stream. The smoke is then illuminated using halogen light sources, and high-resolution cameras are employed to capture the movement of the smoke streaklines over and around the model. Figure 4.15 illustrates the smoke visualization technique applied to the analysis of airflow over a windcatcher prototype in a wind tunnel. This method provides valuable insights into vortex formation, recirculation zones, and flow separation regions, which are otherwise difficult to quantify.

Moreover, such visual observations are often compared with CFD visualizations to evaluate the fidelity of simulation results and to ensure that flow behavior predicted numerically aligns with physical experiments.

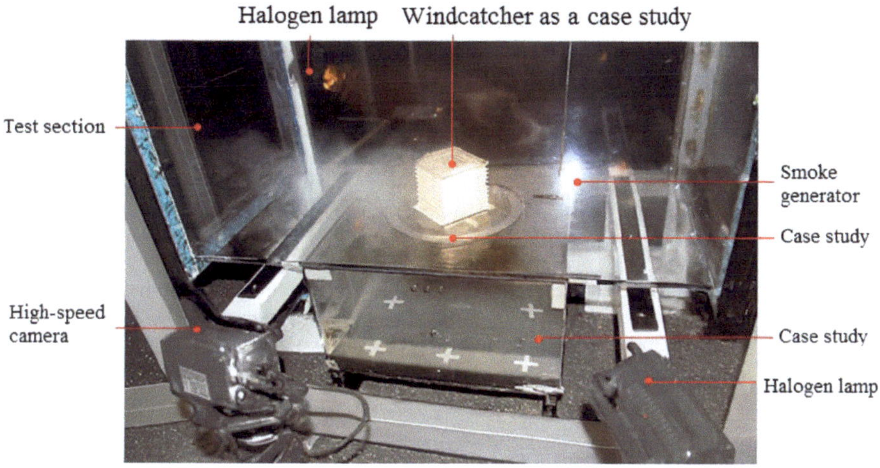

Fig. 4.15 Wind tunnel experiment of a room model with windcatcher: detailed configuration

4.12 Summary

This chapter presents the theoretical foundations and experimental methods required to analyze wind flow around architectural and urban-scale models. It begins with an overview of wind tunnel operation principles, tunnel types, and velocity and turbulence measurement instrumentation. The chapter then details sensor calibration procedures, statistical and spectral data analysis techniques, and the estimation of measurement uncertainty. A novel method is introduced for reconstructing two-dimensional velocity fields using single-axis hot-wire sensors. Experimental results are compared with CFD simulations to evaluate model accuracy. The chapter provides an integrated framework for validating CFD models in climate-responsive building design and urban aerodynamic analysis.

References

1. Zhu, W., & Li, D. (2024). Models for wind tunnel tests based on additive manufacturing technology. *Springer Nature Singapore*. https://doi.org/10.1007/978-981-99-5877-1
2. Shaeri, J., Mahdavinejad, M., & Pourghasemian, M. H. (2022). A new design to create natural ventilation in buildings: Wind chimney. *Journal of Building Engineering*, *59*, 105041. https://doi.org/10.1016/j.jobe.2022.105041
3. Bottasso, C. L., & Campagnolo, F. (2021). Wind tunnel testing of wind turbines and farms. In *Handbook of wind energy aerodynamics* (pp. 1–57).
4. Huang, S., Li, Q. S., & Xu, S. (2007). Numerical evaluation of wind effects on a tall steel building by CFD, *63*, 612–627. https://doi.org/10.1016/j.jcsr.2006.06.033
5. Hufnagel, K. (2022). *Wind tunnel balances*. Springer International Publishing. https://doi.org/10.1007/978-3-030-97766-5

6. Rahmatmand, A., Yaghoubi, M., Rad, E. G., & Tavakol, M. M. (2014). 3D experimental and numerical analysis of wind flow around domed-roof buildings with open and closed apertures. *Building Simulation, 7*(3), 305–319. https://doi.org/10.1007/s12273-013-0157-0
7. Shaeri, J., Mahdavinejad, M., Vakilinejad, R., Bazazzadeh, H., & Monfared, M. (2023). Effects of sea-breeze natural ventilation on thermal comfort in low-rise buildings with diverse atrium roof shapes in BWh regions. *Case Studies in Thermal Engineering, 41*, 102638.
8. Barlow, J. B., Rae, W. H., & Pope, A. (1999). *Low-speed wind tunnel testing*. Wiley.
9. Yang, W., Ying, C.-J., & Yang, Z. (2010). Aerodynamic study of WIG craft near curved ground. *Journal of Hydrodynamics, Series B, 22*(5, Supplement 1), 371–376. https://doi.org/10.1016/S1001-6058(09)60221-3
10. Jiang, Y., & Chen, Q. (2002). Effect of fluctuating wind direction on cross natural ventilation in buildings from large eddy simulation. *Building and Environment, 37*(4), 379–386. https://doi.org/10.1016/S0360-1323(01)00036-1

Chapter 5
CFD Software for Fluid–Thermal Simulation: A Comparative Review for Building and Urban Applications

Abstract This chapter presents a structured comparative review of 24 prominent CFD software platforms, encompassing commercial, open-source, and cloud-based solutions. The analysis covers core capabilities including mesh generation, turbulence modeling, multi-physics coupling, GPU-accelerated solvers, and user interface design. Particular emphasis is placed on their applicability to building physics, architectural aerodynamics, and urban environmental simulation. Tools such as FLUENT, OpenFOAM, COMSOL, and SimScale are assessed for their suitability in modeling complex fluid–thermal interactions in both interior and exterior environments. The chapter aims to support researchers, engineers, and architects in selecting appropriate CFD tools for high-fidelity simulation of ventilation, heat transfer, pollutant dispersion, and wind flow around urban structures. Integration with energy analysis platforms and the growing role of multi-physics simulation in sustainable design are also discussed.

Keywords CFD software · Simulation · Application · Ansys fluent · Comsol

5.1 Introduction

In recent decades, the field of computational fluid dynamics (CFD) has undergone a remarkable transformation, with significant advances in numerical methods, solver algorithms, and software accessibility [1, 2]. A wide spectrum of CFD platforms has emerged, ranging from open-source frameworks designed for academic research to robust commercial suites widely adopted in industry [3, 4]. As CFD continues to expand into interdisciplinary domains, including building physics, environmental design, automotive and aerospace engineering, and urban climate modeling, choosing the appropriate simulation tool has become increasingly critical [5–7].

This chapter provides a systematic and comparative overview of 24 major CFD software packages, encompassing both general-purpose solvers and domain-specific applications. The analysis includes software from leading commercial vendors (e.g., ANSYS Fluent, STAR-CCM +, COMSOL Multiphysics, and NUMECA) as well as

J. Shaeri and A. Cheshmehzangi, *Applied CFD*, Urban Sustainability,
https://doi.org/10.1007/978-981-95-8505-2_5

advanced open-source environments (e.g., OpenFOAM, SU2, and Palabos). Cloud-based and hybrid modeling tools, such as SimScale and DesignBuilder CFD, are also addressed, particularly in light of their increasing role in urban and architectural applications.

For each platform, key features are outlined—including physical modeling capabilities, numerical robustness, mesh generation strategies, solver types, user interface design, and interoperability with other tools. Furthermore, the chapter explores the suitability of each software for applications in architecture, building engineering, urban microclimate simulation, and multi-physics integration. Special focus is placed on tools that enable accurate modeling of airflow, heat transfer, solar radiation, and pollutant dispersion in complex urban geometries.

By consolidating this information in a structured format, this chapter aims to assist CFD practitioners, architects, researchers, and engineers in identifying the most effective simulation tool for their specific design and analysis objectives. It provides a clear comparison of strengths, limitations, and integration potentials across leading platforms, helping users align tool capabilities with project requirements. This guidance is especially valuable in an era of growing demand for accurate, energy-conscious, and climate-responsive environmental modeling.

5.2 Introduction to CFD Softwares

5.2.1 STAR-CCM$^+$ Software

STAR-CCM$^+$ is a widely used commercial CFD software for simulating fluid flow, heat transfer, and reactive phenomena across a range of engineering applications (see Fig. 5.1). It is particularly prevalent in aerodynamic and aerospace engineering, where it has become a standard tool in numerous top-tier universities and industrial research centers. This software offers extensive multi-physics capabilities and a user interface that resembles academic-style workflows, rendering it conceptually similar to ANSYS FLUENT. Another tool from the same family, STAR-CD, is extensively utilized in internal combustion engine simulations.

5.2.2 OPENFOAM Software

OPENFOAM (Open Field Operation and Manipulation) is one of the few open-source CFD platforms, offering unique access to its internal solvers, equations, and modeling techniques. Users can inspect, modify, and extend the source code to implement custom physical models and numerical schemes, making it particularly popular in academic research and open-source development communities.

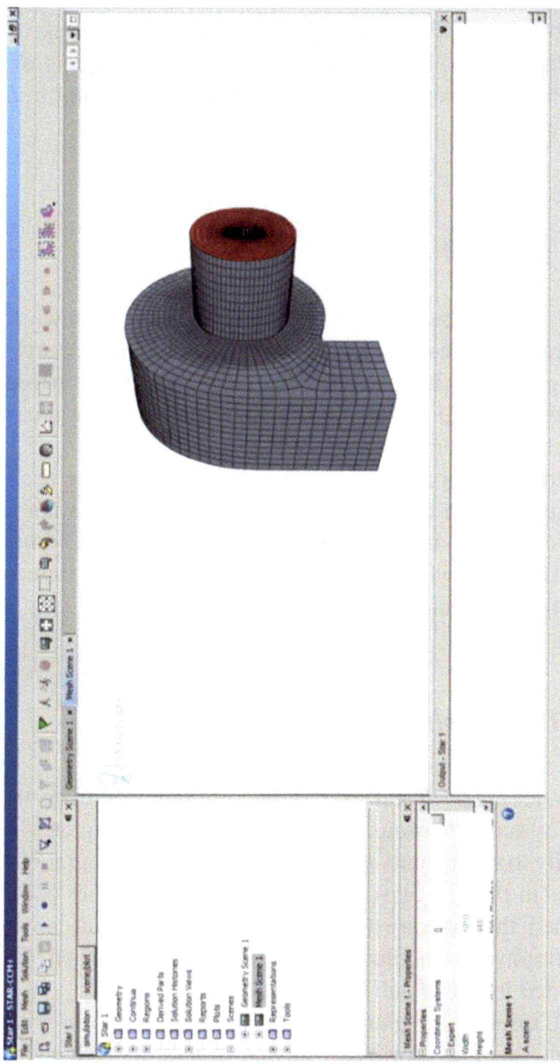

Fig. 5.1 Graphical user interface (GUI) of STAR-CCM + (*Source* Developed by Jalil Shaeri)

OPENFOAM's flexibility has also attracted attention from industry. Some major engineering firms have built specialized CFD tools on top of OPENFOAM's core solvers to address domain-specific challenges. A notable example is SHELL's proprietary explosion modeling software for offshore units and refinery platforms.

Initially developed in the 1980s at Imperial College London, the software is now maintained by the OpenFOAM Foundation, a non-profit organization supported by volunteer contributions from CFD professionals. However, in contrast to commercial tools such as FLUENT and ANSYS CFX, OPENFOAM lacks centralized funding and formal commercial development, resulting in slower evolution and a steeper learning curve for new users.

5.2.3 Flow-3D Software

Flow-3D is another widely used CFD software, particularly known for its robust performance in hydraulic and casting simulations. Although it supports a wide range of physical modeling features, its specialized capabilities in free-surface flows, sediment transport, spillways, and metal solidification make it a preferred choice for civil, water resources, and hydraulic engineers.

The first commercial version of Flow-3D was released in the early 1980s (see Fig. 5.2). Due to its use of non-body-fitted structured meshes, it is not ideally suited for geometries with complex, sharp-edged features. Nevertheless, compared to other CFD tools, Flow-3D is renowned for its ease of use, with an intuitive graphical interface and streamlined simulation workflow.

5.2.4 COMSOL Multiphysics Software

COMSOL Multiphysics is a powerful and flexible simulation platform that has gained increasing popularity across various fields of science and engineering due to its exceptional capability to solve multi-physics problems in an integrated environment. Originally developed in 1986 by two computational fluid dynamics (CFD) specialists in Stockholm, Sweden, the software underwent significant evolution with the release of its first fully multiphysics-enabled version in 1998. Since then, COMSOL has continuously expanded its scope and user base.

One of COMSOL's defining features is its ability to couple a wide range of physical phenomena—such as fluid flow, heat transfer, structural mechanics, electromagnetism, acoustics, and chemical reactions—within a unified computational domain. For instance, it allows the simulation of electromagnetic heating in electric motors, simultaneously solving for airflow, temperature distribution, and electromagnetic fields. This capability makes the software particularly valuable for applications in chemical engineering, where complex transport-reaction systems must be modeled in a coupled manner.

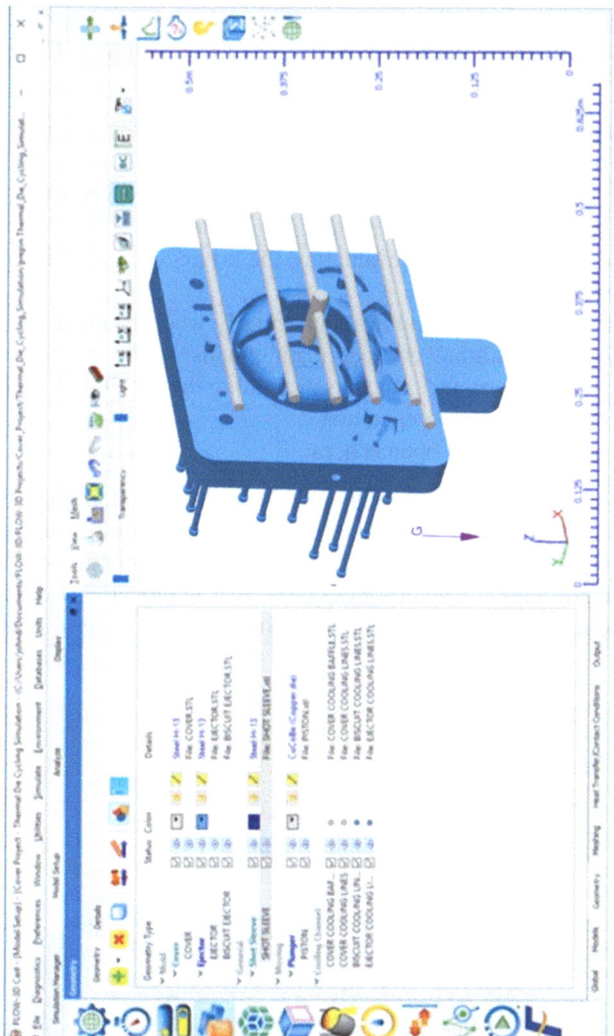

Fig. 5.2 GUI of the Flow-3D software (*Source* Developed by Jalil Shaeri)

Over the last decade, the platform has seen widespread adoption among a broader engineering audience, facilitated by an intuitive graphical user interface (GUI), robust meshing tools, and its equation-based modeling framework that allows users to directly edit or introduce custom governing equations.

Beyond its traditional use in engineering disciplines, COMSOL Multiphysics has become increasingly relevant in the fields of architecture, urban design, and environmental planning, especially as sustainable and performance-based design strategies have gained traction. In these contexts, CFD simulations in COMSOL are used to evaluate:

- Thermal comfort in indoor and outdoor spaces under varying solar radiation and ventilation conditions,
- Wind flow analysis around complex building forms and urban canyons to assess pedestrian wind safety and natural ventilation potential,
- Pollutant dispersion modeling for evaluating the urban air quality in residential and traffic-heavy zones,
- Daylighting and solar radiation studies, particularly when coupled with thermal performance models to estimate cooling/heating loads,
- Green infrastructure performance, such as evaluating the cooling impact of green roofs, green facades, and urban vegetation,
- Acoustic propagation in urban open spaces.

The multi-physics capabilities of COMSOL are essential when addressing built environment problems involving the coupling of radiation, convection, conduction, and fluid movement (see Fig. 5.3). For example, when modeling the microclimate of an urban courtyard, the software can incorporate convective air movement, solar loading, surface heat absorption, and even moisture transport in soil or plant systems.

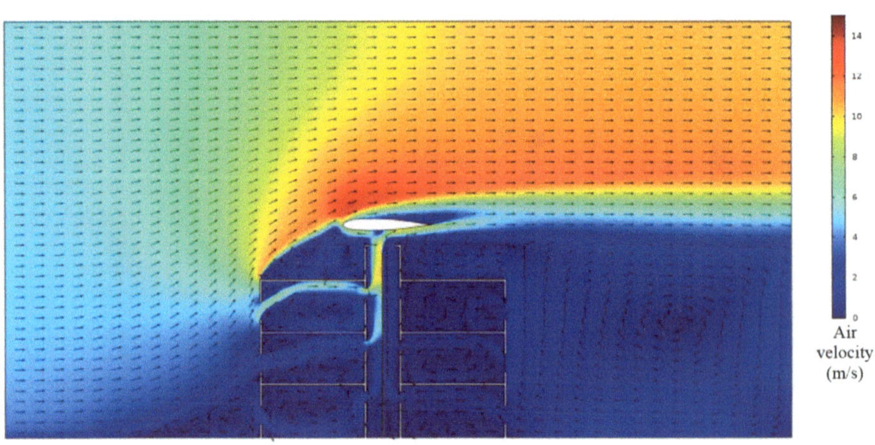

Fig. 5.3 GUI of the COMSOL software

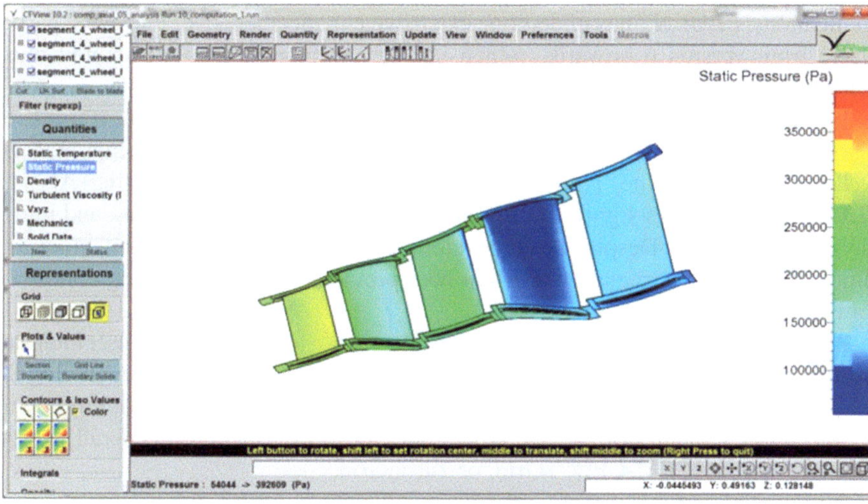

Fig. 5.4 GUI of the NUMECA software (*Source* Developed by Jalil Shaeri)

Thanks to its extensibility, COMSOL also allows direct linking with climate datasets, BIM models, and Python/Matlab scripts for parametric studies or optimization-based simulations, which are increasingly demanded in performance-driven architectural design.

5.2.5 NUMECA Software

NUMECA is a well-established CFD software suite that has recently experienced a notable growth in popularity and credibility, particularly within the turbomachinery industry (see Fig. 5.4). Its key advantage is the ability to quickly and efficiently generate computational grids and perform simulations for rotating machinery components, such as compressors and turbines. NUMECA's tools facilitate the simultaneous simulation of multiple blade rows—a critical requirement in the design and optimization of modern turbomachinery. Owing to this capability, the software has gained traction across various industrial sectors engaged in high-speed rotating equipment.

5.2.6 ANSYS FLUENT Software

ANSYS FLUENT is one of the most widely used general-purpose commercial CFD packages, with its origins dating back to 1983, alongside a contemporary tool known as PHOENICS. Initially developed for structured mesh solvers and limited academic applications, FLUENT has undergone extensive enhancement over the years, particularly from version 6 onwards, which introduced support for unstructured mesh generation.

In 2006, the software was acquired by ANSYS Inc., and since then, it has been marketed under the name ANSYS FLUENT, integrated into the comprehensive ANSYS simulation suite. The most recent release at the time of this writing, ANSYS FLUENT 16, features significant advancements in turbulence modeling and multi-physics integration. Despite these developments, the core solver architecture and simulation workflow remain consistent with previous versions, ensuring continuity for experienced users.

A notable change in version 16 is the introduction of a tree-structured user interface for simulation settings, streamlining access to model parameters without altering the fundamental process. FLUENT is based on the finite volume method, employing cell-centered variable storage, and typically requires external mesh generation tools such as Gambit, ICEM-CFD, TGrid, or TurboGrid.

The typical computational workflow in ANSYS FLUENT follows a structured sequence comprising three primary stages: mesh generation, simulation setup and execution, and post-processing. The process begins with "mesh generation," where the computational domain is discretized into a finite number of control volumes. This mesh can be created using dedicated pre-processing tools such as Gambit, ICEM-CFD, or other mesh generators compatible with FLUENT. Once generated, the mesh is imported into FLUENT in the.msh file format, ready for the application of physical models and numerical settings. Next is the "simulation setup and execution" stage. Within the FLUENT environment, users define the governing physical models (e.g., turbulence, heat transfer, and multiphase flow), assign appropriate boundary and initial conditions, and configure solver parameters. Upon execution, the solver iterates through the defined equations, and simulation results are stored in the.dat file format, which records cell-wise values of relevant field variables such as velocity, pressure, and temperature. Finally, the "post-processing" phase allows users to interpret and analyze the simulation output. Visualization of results can be carried out directly within FLUENT using its built-in graphical tools. Alternatively, data can be exported to ANSYS CFD-Post, a versatile post-processing platform shared with ANSYS CFX. CFD-Post provides advanced visualization capabilities, including vector plots, contour maps, streamlines, and quantitative data extraction, enabling a detailed and insightful evaluation of simulation outcomes.

Recent releases of ANSYS-FLUENT, particularly those introduced after 2020, have incorporated substantial structural and functional enhancements aimed at improving user experience and increasing the fidelity of numerical modeling. One of the key developments is the integration of a modern workflow-based graphical user interface (GUI), which streamlines the simulation process. Additionally, full interoperability with the ANSYS Workbench environment and significant upgrades to the internal meshing tools—most notably the Watertight Geometry Meshing capability—have markedly enhanced usability and mesh quality.

These versions also feature adaptive mesh refinement (AMR), advanced multiphase flow modeling, and improved accuracy in the simulation of combustion, mass

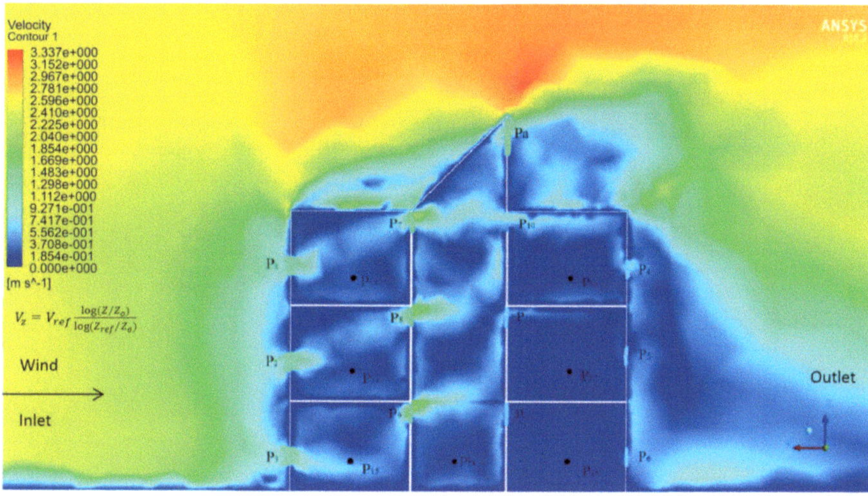

Fig. 5.5 Graphical user interface (GUI) of the FLUENT software (*Source* Developed by Jalil Shaeri)

transfer, and complex chemical reactions. The incorporation of GPU-accelerated solvers has led to notable improvements in computational efficiency, particularly for large-scale CFD problems. Furthermore, the inclusion of advanced turbulence models such as the SST Transition and k–ω BSL, along with the discrete phase model (DPM) and high-fidelity thermal modeling capabilities, make the recent versions of ANSYS-FLUENT more suitable than ever for a wide range of engineering applications—including building physics, energy systems, automotive design, and aerospace engineering (see Fig. 5.5).

5.2.7 CFX Software

CFX is one of the relatively long-standing computational fluid dynamics (CFD) tools that have gained wide adoption due to its powerful capabilities and user-friendly interface. The initial release of this software dates back to 1996. Subsequently, through the acquisition of TASCflow, it extended its capability to perform simulations on unstructured grids. Originally developed for nuclear physics applications, CFX is currently part of the ANSYS product suite. It employs a finite-volume approach combined with finite-element interpolation, storing variables at nodal locations.

CFX is particularly effective in simulating turbomachinery and aerodynamic flows, often producing highly accurate results with relatively short turnaround times. Owing to its use of a coupled solver and nodal-based storage, it requires greater memory compared to ANSYS FLUENT. However, its implementation of a pseudo-transient coupled solution strategy for steady-state simulations results in significantly

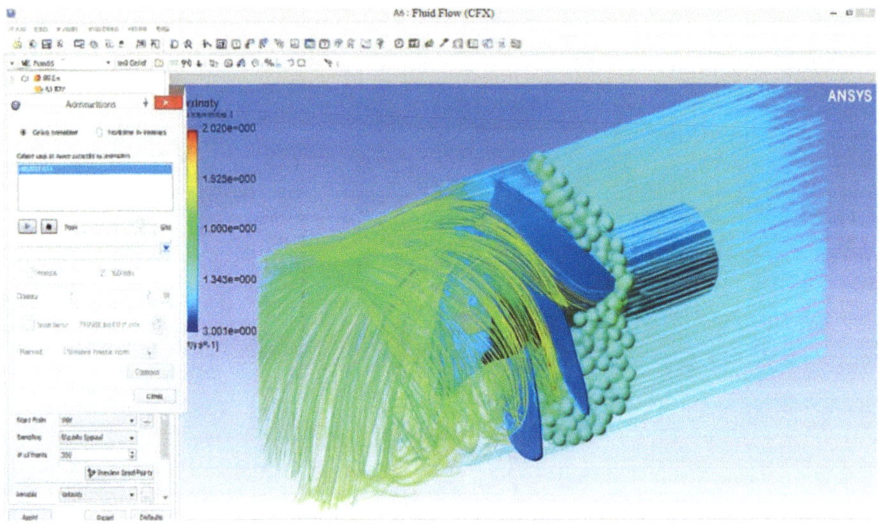

Fig. 5.6 Graphical user interface (GUI) of the CFX software (*Source* Developed by Jalil Shaeri)

faster convergence than FLUENT. Although this pseudo-transient scheme was later integrated into FLUENT from version 14 onward, it appears to offer less robust convergence behavior compared to CFX.

In terms of turbulence modeling, CFX slightly outpaces FLUENT in the breadth of available models (see Figs. 5.6 and 5.7). Nevertheless, for specialized physics such as flows involving moving boundaries or reactive systems, FLUENT generally provides more advanced modeling capabilities. It is worth noting that the primary user base for CFX is currently industrial, and it tends to be favored over FLUENT in sectors such as rotating machinery design. According to unofficial statements made by ANSYS during a webinar, a unified solver combining CFX and FLUENT was planned for post-version 15 releases. However, as of version 16, this integration had not materialized.

5.2.8 *GAMBIT Software*

GAMBIT is one of the earliest dedicated meshing tools developed for CFD applications. Originally designed for Unix/Linux platforms, it requires the Exceed interface to run on Windows environments. Following ANSYS's acquisition of the FLUENT suite, the development and distribution of GAMBIT were discontinued, with no new versions released after 2006.

Despite its discontinuation, GAMBIT continues to be widely used in both academic and industrial settings, particularly in countries where legacy systems are still prevalent. For this reason, selected sections of this book retain references

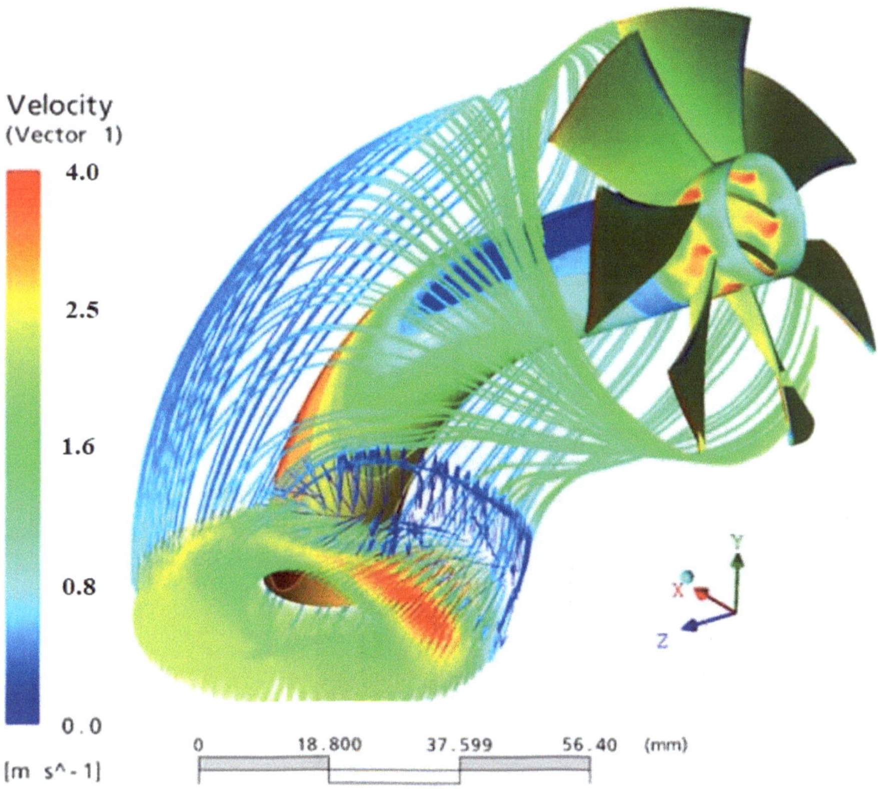

Fig. 5.7 Post-processed simulation output from the CFX software (*Source* Developed by Jalil Shaeri)

to GAMBIT-based workflows, although these will be gradually revised in future editions using alternative meshing tools. GAMBIT excels in generating unstructured surface meshes using Delaunay triangulation and, in certain aspects, offers advantages over ICEM-CFD for this specific task.

Although no longer officially supported, the final version of GAMBIT is made available for educational purposes on the PITEC Engineering website. Moreover, for users who prefer not to learn GAMBIT (see Fig. 5.8), pre-generated mesh files corresponding to Chaps. 2 and 3 are provided online.

5.2.9 ICEM-CFD Software

As a core component of the ANSYS suite, ICEM-CFD is among the most powerful and widely used mesh generation tools available today. It is extensively adopted in both academic and industrial CFD projects. ICEM-CFD supports the generation of

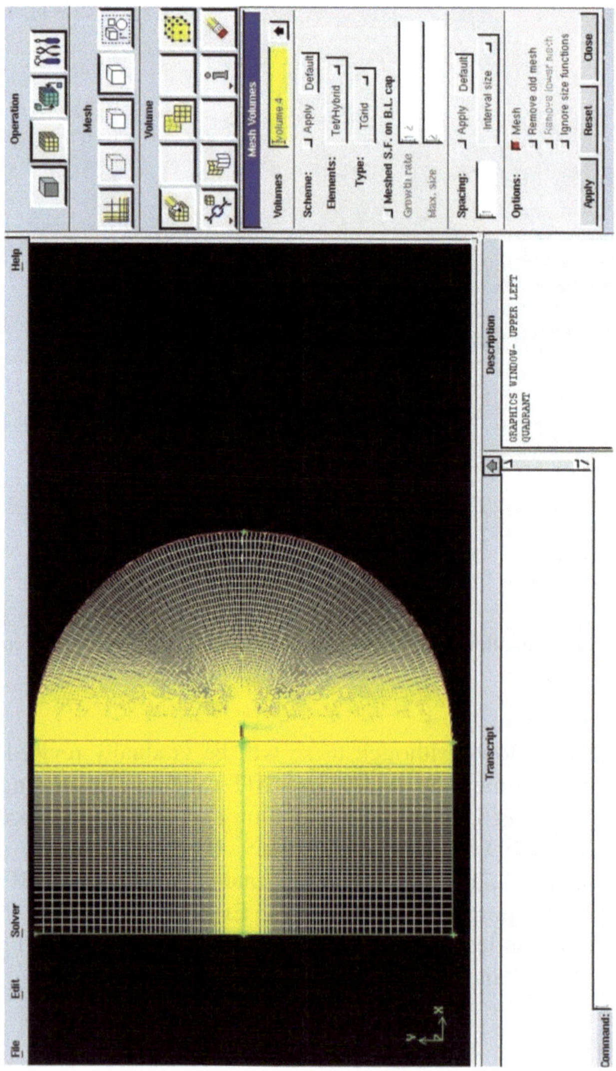

Fig. 5.8 Graphical user interface (GUI) of the GAMBIT mesh generator (*Source* Developed by Jalil Shaeri)

both structured and unstructured meshes and offers robust integration with a wide range of CAD tools such as CATIA and SolidWorks. These capabilities make it a versatile and essential utility for complex simulations.

Major global organizations—including General Electric, NASA, Mercedes-Benz, Ford, and Formula 1 teams—are among the notable users of ICEM-CFD. While ICEM-CFD excels in mesh generation, its geometry creation and editing tools are relatively limited. However, the recent acquisition of SpaceClaim by ANSYS and its seamless integration with ICEM-CFD has significantly mitigated these limitations, providing users with an efficient geometry preparation workflow (see Fig. 5.9).

5.2.10 TGrid Software

TGrid, currently integrated under the name FLUENT-Meshing, is a powerful tool for generating unstructured computational grids. It is installed alongside the FLUENT package and can be accessed directly through the FLUENT launch interface. This software offers a relatively user-friendly interface that enables the rapid generation of unstructured and boundary-layer meshes over complex three-dimensional geometries.

Recent versions of TGrid have incorporated some of the meshing capabilities originally available only in ICEM-CFD. Although it does not support geometry creation natively, since version 14.5 a geometric interface has been added, allowing users to import geometries from popular CAD tools (see Figs. 5.10 and 5.11). Furthermore, version 16 includes enhanced features for geometry repair and modification (see Figs. 5.12 and 5.13).

Due to significant changes in user interface and internal structure starting from version 16, it is recommended to use versions 14, 14.5, or 15 for educational purposes and for compatibility with the tutorial content in this book.

5.2.11 CFD-POST Software

Originally developed as a post-processing utility for ANSYS CFX, the CFD-POST software has evolved into the primary post-processor for both CFX and FLUENT within the ANSYS ecosystem. Unlike FLUENT's built-in post-processing tools, CFD-POST offers high-quality visualization, advanced rendering effects, and support for generating HD animations with customizable frame rates.

Given its growing adoption in both academic and industrial settings, sections of this book have been tailored to incorporate analysis and visualization using CFD-POST (see Fig. 5.14). This software is bundled with the ANSYS suite and is easily accessible following installation.

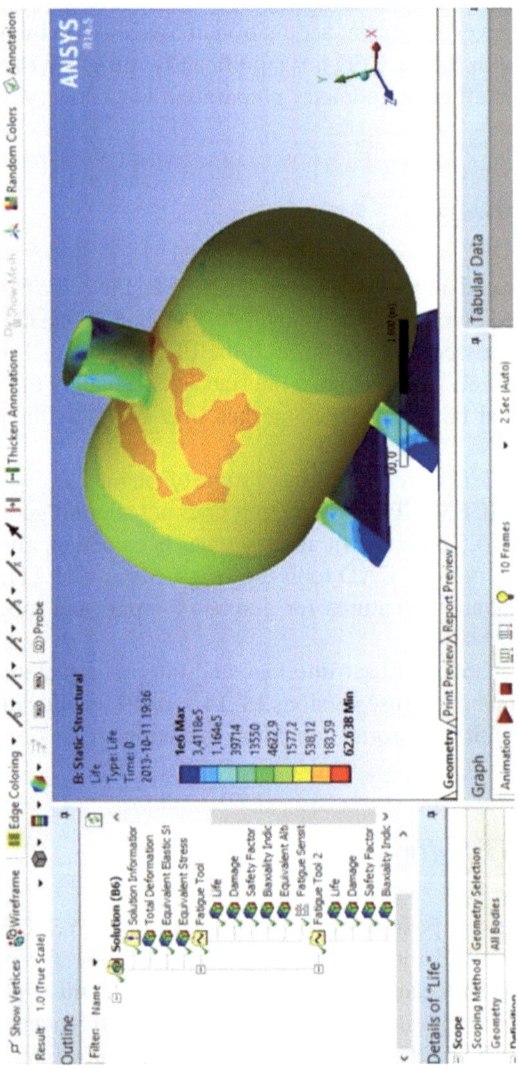

Fig. 5.9 Graphical user interface (GUI) of the ICEM-CFD meshing tool (*Source* Developed by Jalil Shaeri)

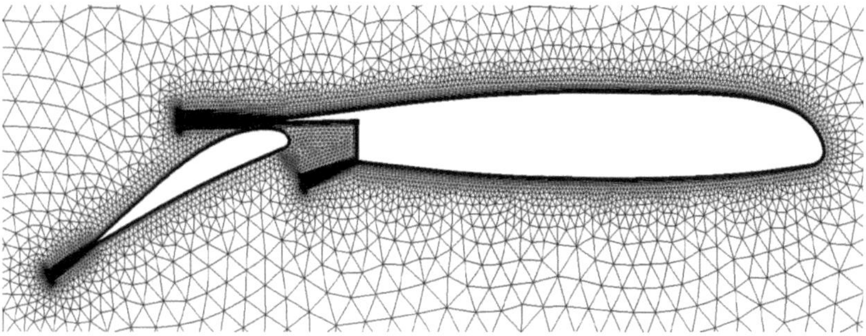

Fig. 5.10 Two-dimensional mesh generated in TGrid software (*Source* Developed by Jalil Shaeri)

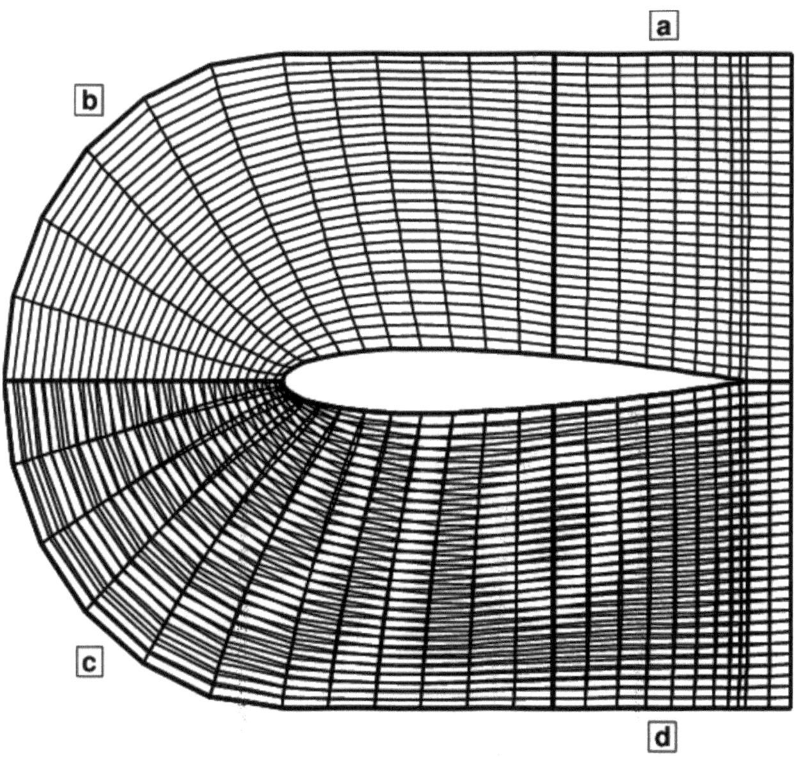

Fig. 5.11 Two-dimensional mesh in TGrid showing boundary layer refinement (*Source* Developed by Jalil Shaeri)

5.2.12 AVL FIRE Software

AVL FIRE is one of the flagship simulation tools developed by AVL, specifically designed for heat transfer and fluid flow simulations in internal combustion engines.

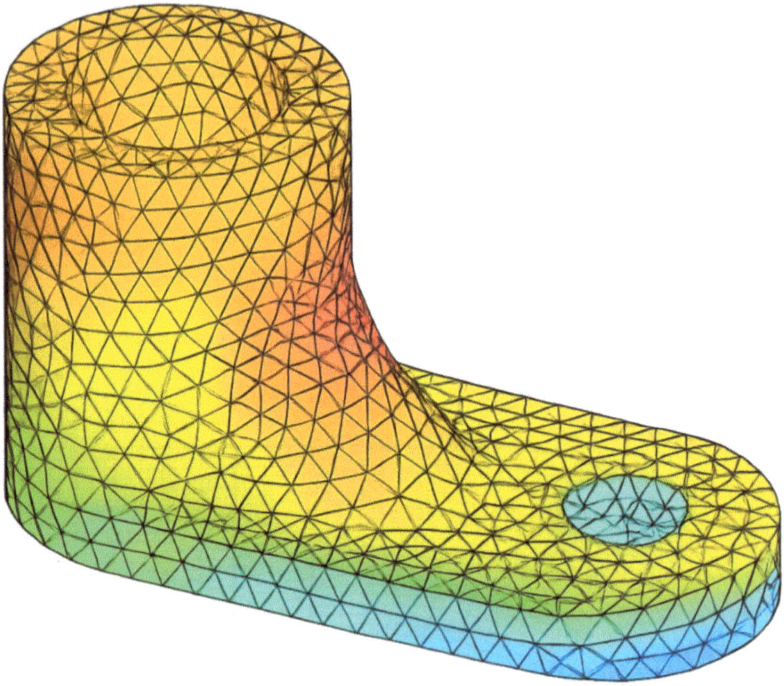

Fig. 5.12 Three-dimensional mesh generated in TGrid software (*Source* Developed by Jalil Shaeri)

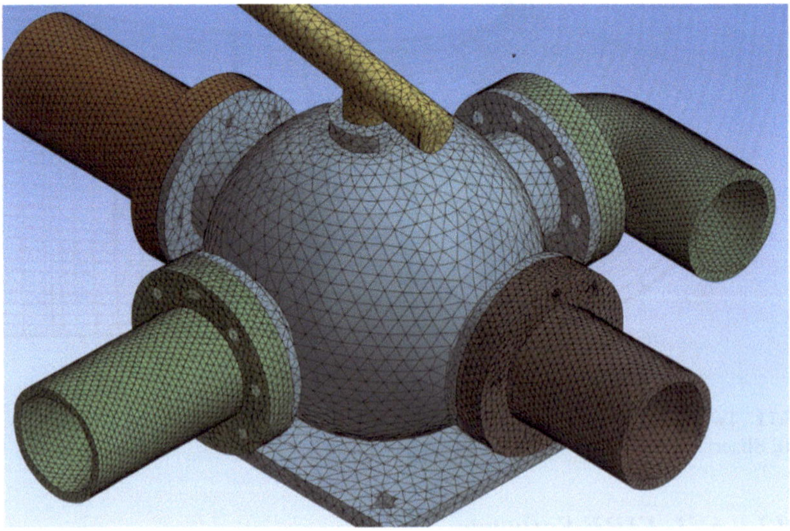

Fig. 5.13 Three-dimensional structured/unstructured mesh in TGrid software (*Source* Developed by Jalil Shaeri)

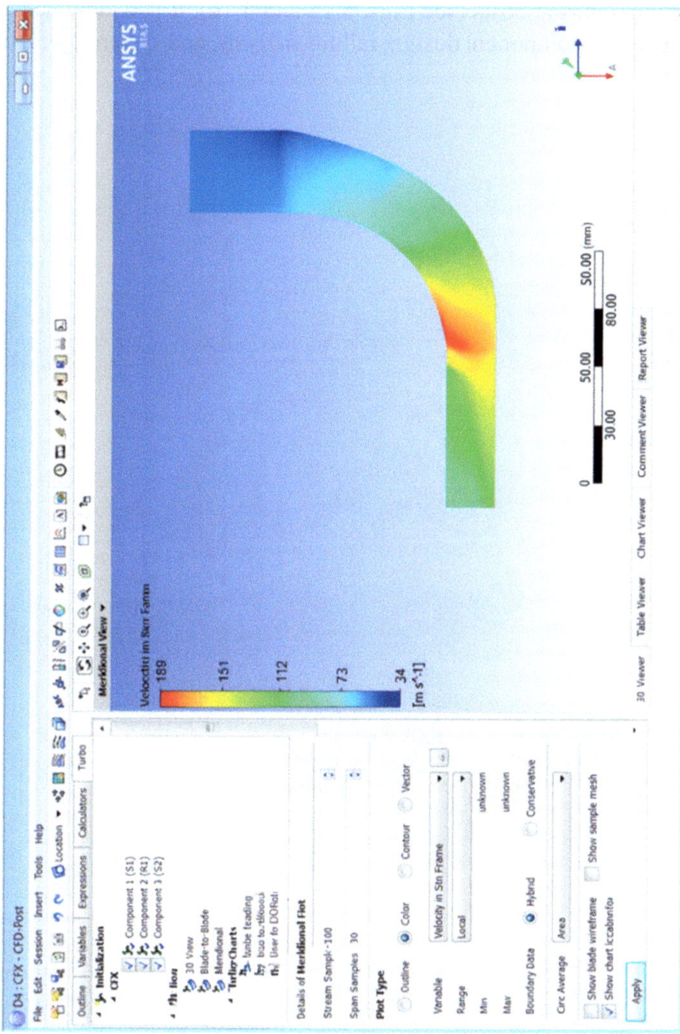

Fig. 5.14 Graphical user interface (GUI) of the CFD-POST post-processing tool (*Source* Developed by Jalil Shaeri)

It provides highly accurate modeling capabilities for engine thermal behavior and fluid dynamics. The software is capable of solving governing flow equations and predicting complex phenomena such as in-cylinder heat transfer, multiphase flows, and dynamic thermal loading.

AVL FIRE enables automotive engineers to extract detailed thermal insights and performance metrics, offering predictive capabilities for optimizing engine lifespan and reducing manufacturing costs (see Figs. 5.15 and 5.16). It is widely used in the automotive industry for component design, failure analysis, and system-level thermal management.

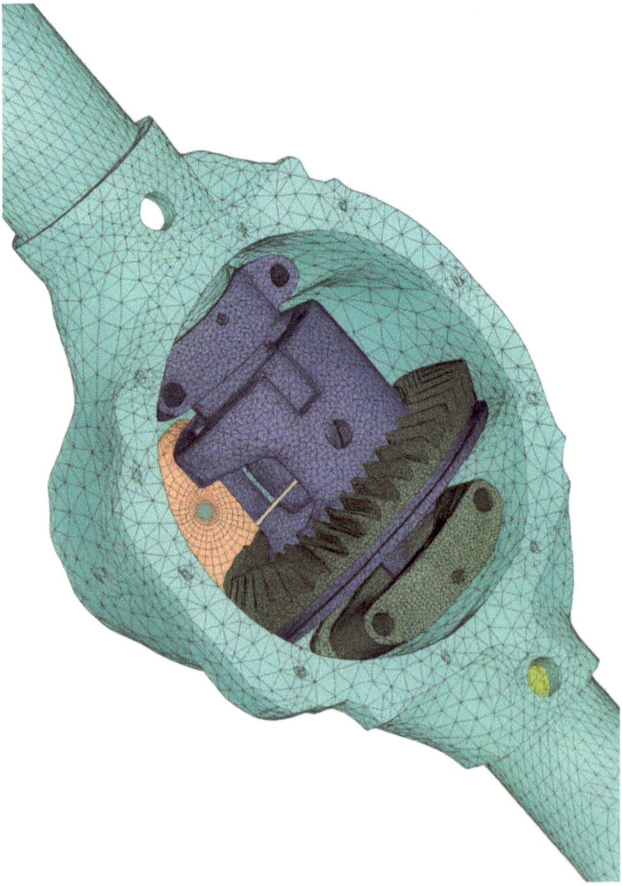

Fig. 5.15 Mesh generation in AVL Fire for internal combustion engine modeling software (*Source* Developed by Jalil Shaeri)

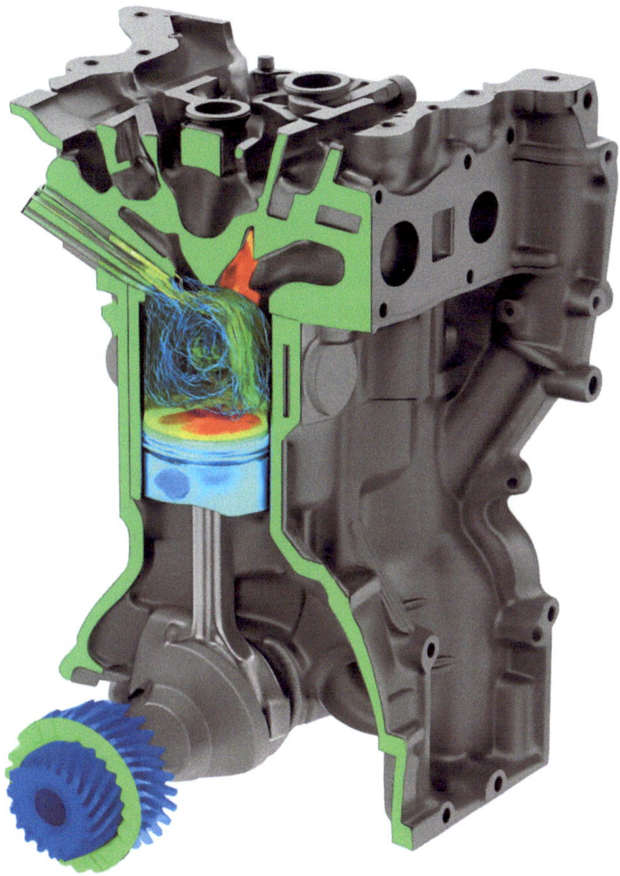

Fig. 5.16 Simulation results output from AVL Fire (*Source* Developed by Jalil Shaeri)

5.2.13 CFD++ Software

CFD++ is a high-end computational fluid dynamics software package developed for simulating compressible and incompressible, steady and unsteady flows across a broad range of Mach numbers, including subsonic, sonic, supersonic, and hypersonic regimes. It supports the modeling of laminar and turbulent flows, multiphase effects, cavitation, and radiative heat transfer phenomena.

A major strength of CFD++ lies in its advanced meshing capabilities, allowing users to efficiently handle complex geometries using a hybrid mesh approach—combining structured, unstructured, and multiblock grids. The solver supports skewed and non-conformal meshes and maintains numerical stability and accuracy in challenging simulation environments.

CFD++ is particularly well-suited for applications involving moving and stationary bodies, and it has been successfully employed in aerospace, propulsion, turbomachinery, and advanced thermal systems.

5.2.14 PumpLinx Software

PumpLinx is an advanced three-dimensional computational fluid dynamics (CFD) tool designed specifically for the virtual testing, analysis, and performance prediction of liquid-handling systems. These include pumps, motors, compressors, turbines, valves, propellers, and other hydraulic or rotating equipment involving complex fluid–structure interactions.

PumpLinx is particularly well-suited for fluid systems and features a proprietary cavitation module that enables highly accurate modeling of vapor formation, gas release, and liquid compressibility. This allows for detailed assessment of performance degradation, pressure surge phenomena, and cavitation-induced damage under real-world operating conditions.

5.2.15 Autodesk CFD

Autodesk CFD is a flexible simulation platform for computational fluid dynamics and thermal analysis. It is engineered to provide engineers with an integrated environment to explore, visualize, and optimize the thermal and flow behavior of products prior to manufacturing.

The primary objective of Autodesk CFD is to simulate design performance, optimize configurations, and validate thermal-fluid behavior early in the product development cycle. This reduces reliance on physical prototyping, minimizes trial-and-error testing, and significantly cuts design costs and lead times. It is widely used in mechanical design, electronics cooling, and HVAC system simulation.

5.2.16 DesignBuilder CFD Module

DesignBuilder CFD is a powerful simulation module embedded within the DesignBuilder platform, primarily used for building performance modeling. It supports dynamic modeling of various aspects of a building—including envelope materials, architecture, HVAC systems, lighting, and energy usage patterns.

Beyond heating and cooling load estimation, DesignBuilder simulates detailed energy consumption components such as lighting demand, plug loads, domestic hot water, and more. Its CFD module enables both indoor and outdoor airflow simulation. The indoor CFD solver is used for assessing thermal comfort, ventilation performance, and HVAC system behavior, while the outdoor module evaluates the influence of wind flow on building form and urban morphology. DesignBuilder utilizes local climatic data from official weather files to accurately simulate site-specific energy behavior and environmental performance (see Figs. 5.17 and 5.18).

5.2.17 ENVI-Met Urban CFD Platform

ENVI-met is a specialized CFD-based microclimate modeling tool developed for urban designers, landscape architects, and urban planners. It simulates surface–atmosphere interactions at the neighborhood scale by incorporating a wide range of urban materials and green infrastructure elements—such as walls, pavements, vegetation, green roofs, and living walls.

ENVI-met enables high-resolution simulation of key environmental parameters, including air temperature, relative humidity, wind velocity, and outdoor thermal comfort indices. One of its core capabilities is urban CFD analysis, allowing users to evaluate airflow and pollutant dispersion within complex urban morphologies. It is widely adopted for sustainable urban design, passive cooling assessment, and urban heat island mitigation strategies (see Fig. 5.19).

5.2.18 SimScale Software

SimScale is an innovative platform for computational fluid dynamics (CFD) and multiphysics simulation, fully based on cloud computing technology. It enables users to perform complex simulations involving fluid flow, heat transfer, structural mechanics, and aerodynamics directly through a web browser, eliminating the need for local high-performance computing resources. The backend solvers of SimScale are built upon open-source frameworks such as OpenFOAM and Code_Aster, allowing for robust and scalable computations. Furthermore, its tight integration with CAD environments and a modern, workflow-oriented graphical user interface makes it a compelling option for both intermediate and advanced users. The platform's on-demand cloud infrastructure facilitates rapid execution of large-scale, high-fidelity simulations without local computational limitations.

SimScale finds application across a wide range of domains, including airflow analysis in and around buildings, HVAC system design, external vehicle aerodynamics, electronics cooling, urban wind comfort studies, and thermal comfort assessment in interior spaces. One of its standout features is support for incompressible and transient flow modeling, conjugate heat transfer (CHT), multi-phase simulations, and outdoor environmental simulations tailored to pedestrian comfort assessments. Thanks to its cloud-native architecture, users can run multiple concurrent simulations

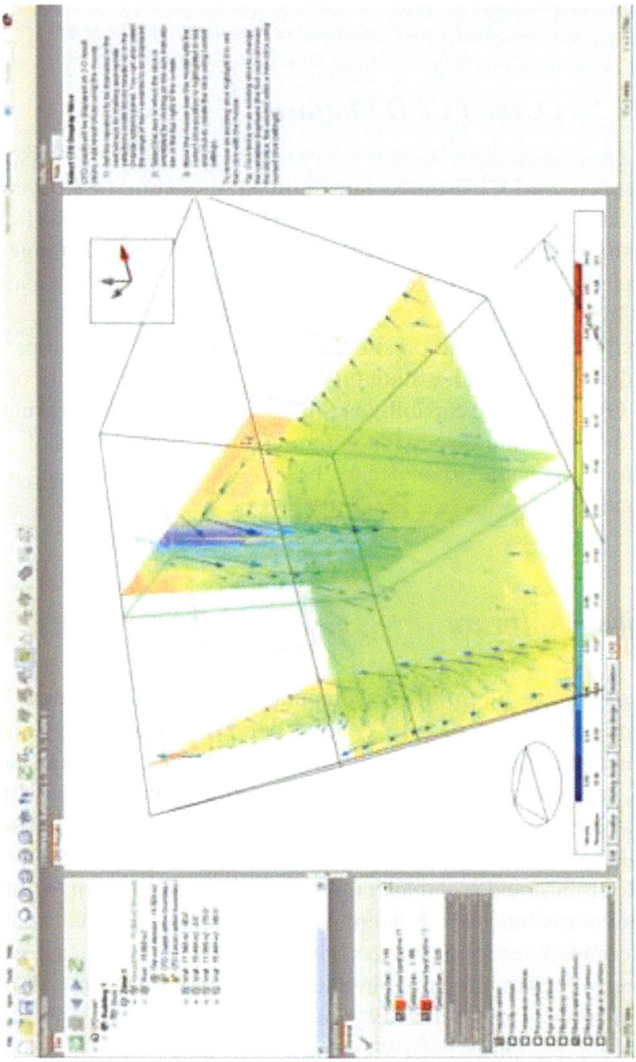

Fig. 5.17 Graphical user interface (GUI) of the DesignBuilder software (*Source* Developed by Jalil Shaeri)

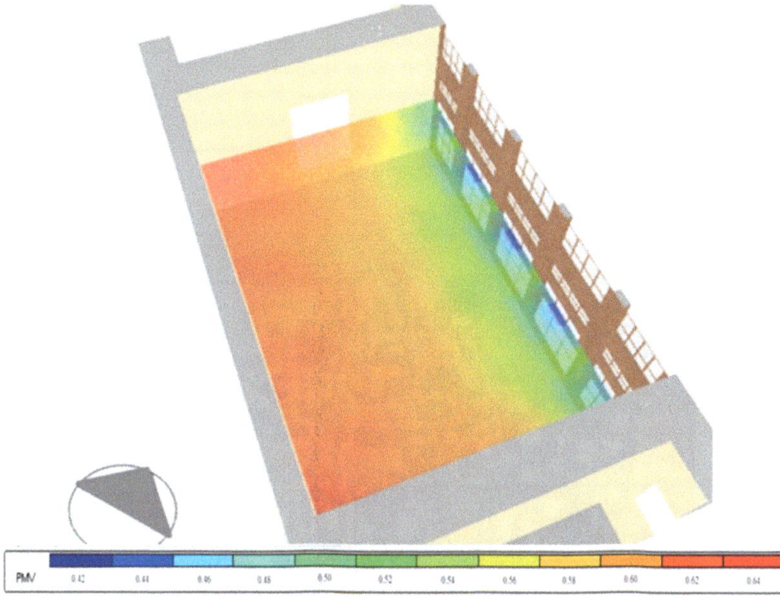

Fig. 5.18 CFD simulation results output from DesignBuilder (*Source* Developed by Jalil Shaeri)

with high efficiency. The combination of advanced numerical models, high accessibility, rapid solution times, and browser-based deployment has positioned SimScale as a modern and powerful CFD tool for professionals in architecture, urban planning, environmental engineering, and mechanical system design.

5.2.19 *XFlow (by Dassault Systèmes)*

XFlow is an advanced computational fluid dynamics (CFD) software developed by Dassault Systèmes, distinguished by its utilization of the lattice Boltzmann method (LBM) as its core numerical approach. Unlike conventional CFD solvers based on the Navier–Stokes equations and structured meshes, XFlow employs a meshless, particle-based formulation that allows for highly automated preprocessing and seamless handling of complex geometries. Its native ability to model transient, turbulent, and multiphase flows without the need for extensive mesh generation makes it especially suitable for fast-paced industrial design processes, where geometry complexity and time constraints are critical considerations. Moreover, XFlow integrates well within the 3DEXPERIENCE platform, enhancing its compatibility with other Dassault Systèmes design and simulation tools.

XFlow is particularly effective in industries such as aerospace, automotive, defense, and marine engineering, where accurate prediction of unsteady flow phenomena—such as vortex shedding, sloshing, or aerodynamic noise—is essential.

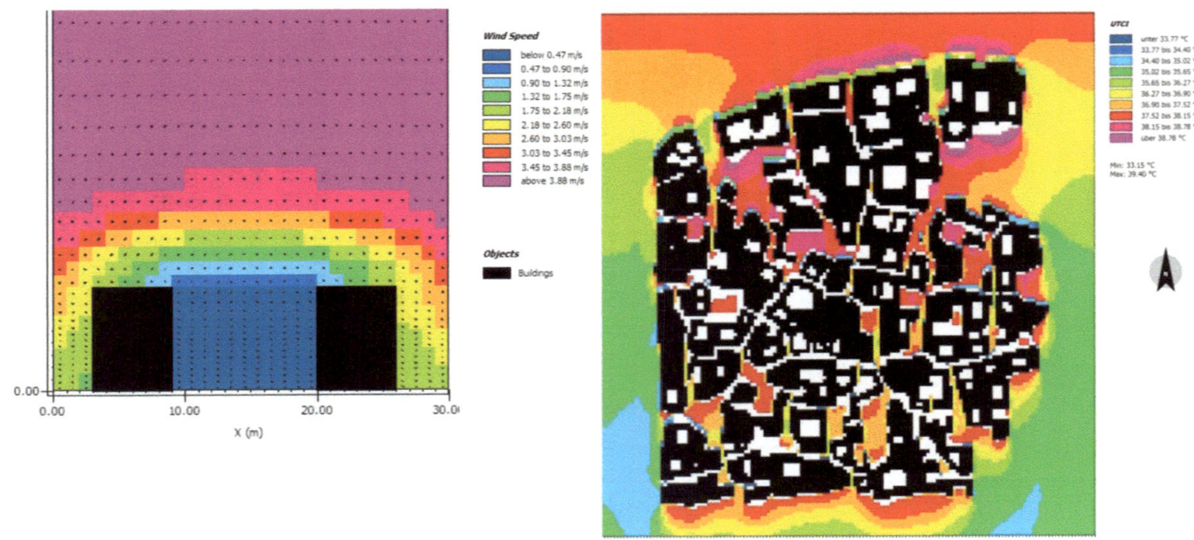

Fig. 5.19 GUI of the ENVI-met urban microclimate model (*Source* Developed by Jalil Shaeri)

Its key strengths lie in its robust turbulence modeling capabilities (including LES), its strong support for fluid–structure interaction (FSI), and its ability to handle moving geometries with high fidelity. Engineers benefit from a simplified user interface, efficient parallel computing, and the reduced need for manual meshing and boundary condition tuning. These features make XFlow an ideal tool for early-stage design evaluation, rapid prototyping, and real-time simulation feedback, contributing significantly to accelerated product development cycles in multidisciplinary engineering environments.

5.2.20 *Su2*

SU2 (Stanford University Unstructured) is an open-source, high-fidelity simulation suite developed initially at Stanford University and now maintained by a global community of contributors. Designed for solving a wide range of partial differential equations (PDEs), SU2 is primarily focused on computational fluid dynamics (CFD), with advanced modules for aerodynamic analysis, adjoint-based shape optimization, multiphysics coupling, and turbulence modeling. Written in C++, SU2 supports unstructured meshes and incorporates modern numerical methods such as finite-volume and discontinuous Galerkin discretizations, making it highly flexible and extensible for research and industrial use alike.

In terms of domain applications, SU2 has been widely adopted in the aerospace, automotive, and energy sectors for tasks such as compressible and incompressible flow simulation, high-speed aerodynamics, shape optimization, and transonic flow prediction. The software includes built-in support for Reynolds-Averaged Navier–Stokes (RANS) turbulence models, adjoint solvers for sensitivity analysis, and time-dependent simulations. Its robust command-line interface and Python-based scripting tools enable easy integration into automated design workflows and high-performance computing clusters.

In the context of architecture and urban planning, SU2's capabilities can be leveraged for simulating urban airflow, thermal comfort analysis, and natural ventilation studies. Although SU2 is less commonly used in architectural practices compared to GUI-based software, its open-source nature and customizable physics solvers make it particularly appealing for academic research, especially in parametric urban design, passive cooling strategies, and wind comfort evaluations in complex urban environments. Researchers can embed SU2 into parametric design platforms or use it in conjunction with mesh generators like Gmsh to simulate airflow around buildings or within urban canyons, thus supporting sustainable urban development initiatives.

5.2.21 HiFUN (High-Fidelity Unstructured Navier–Stokes Solver)

HiFUN is a high-performance CFD solver developed by Hindustan Aeronautics Ltd. (HAL) and utilized extensively in the Indian aerospace and defense sectors. It is designed for the simulation of compressible and incompressible flows over complex geometries using unstructured grids. The software supports steady and unsteady flows and includes a range of turbulence models (such as Spalart–Allmaras and k–ω SST), high-resolution numerical schemes, and parallel computing capabilities based on domain decomposition techniques. HiFUN is optimized for aerospace applications including aircraft aerodynamics, missile design, and re-entry vehicles, but is also increasingly used in academia for research into transonic and supersonic flow regimes. Although its application in architecture or urban design is limited due to its specialization in high-speed external flows, HiFUN's modular structure and open-ended solver configuration may offer potential for academic investigations in urban aerodynamics where compressibility effects are significant.

5.2.22 Palabos (Parallel Lattice Boltzmann Solver)

Palabos is an open-source CFD framework based on the lattice Boltzmann method (LBM), particularly well-suited for simulating complex flow phenomena in domains with irregular geometries and multiphase interactions. Developed in C++ and optimized for parallel execution using MPI, Palabos supports a wide range of physical models, including porous media, non-Newtonian flows, fluid–structure interaction (FSI), and multiphase dynamics. It is widely used in both industrial and academic research, especially in biomedical engineering, microfluidics, and environmental flow modeling. In the domain of architecture and urban planning, Palabos is increasingly recognized for its ability to simulate pedestrian-level wind environments, air quality distribution, and natural ventilation in urban canyons. Its efficient parallel scalability and ease of integration with external meshing and visualization tools make it a viable platform for high-resolution simulations of urban microclimates, thermal comfort, and pollutant dispersion in dense cityscapes.

5.2.23 PowerFLOW (by Exa Corporation, Now Part of Dassault Systèmes)

PowerFLOW is a highly specialized, commercial CFD software platform developed originally by Exa Corporation and now part of the Dassault Systèmes SIMULIA suite. Unlike conventional Navier–Stokes solvers, PowerFLOW is based on the lattice Boltzmann method (LBM), which enables inherently transient, compressible, and scalable simulations using voxel-based meshing. This framework makes PowerFLOW particularly suited for aeroacoustics, external aerodynamics, and thermal management in the automotive, aerospace, and high-performance sports sectors. The software is capable of simulating transient phenomena such as wind noise, cooling airflow, and drag reduction strategies with high accuracy, even over complex geometries without requiring extensive mesh generation. While PowerFLOW has not been widely adopted in architecture or urban design, its strengths in noise propagation modeling, vehicular aerodynamics in urban traffic flows, and passenger thermal comfort offer potential extensions to urban mobility, roadway design, and outdoor wind environment studies, particularly in smart city planning and noise-sensitive zones.

5.2.24 Phoenix Integration (ModelCenter)

Phoenix Integration, through its flagship product ModelCenter, provides a comprehensive platform for workflow integration, process automation, and design optimization across multiple engineering disciplines, including CFD. While ModelCenter is not a CFD solver in itself, it serves as a central coordination hub, linking different simulation tools (such as ANSYS Fluent, STAR-CCM + , and OpenFOAM) with CAD, finite element, and control system models. Its primary advantage lies in enabling multidisciplinary design optimization (MDO), sensitivity analysis, and parametric studies through a graphical user interface that orchestrates diverse simulation codes. ModelCenter is extensively used in aerospace, automotive, and defense sectors, where system-level performance must be optimized across interacting subsystems. In the context of architectural engineering and urban simulation, Phoenix Integration can facilitate the coupling of energy simulation tools, CFD-based airflow analyses, and lighting/daylighting models into a unified workflow. This makes it particularly valuable in integrated building performance simulation, zero-energy building design, and urban sustainability studies, where trade-offs among energy, airflow, and structural systems must be systematically evaluated.

5.3 Comparative Analysis of Twenty-Four CFD Software Platforms for Fifteen Key Architectural and Urban Applications

The integration of CFD into the fields of architecture and urban planning has become a cornerstone of performance-based design, sustainable development, and environmental resilience [3, 8, 9]. By enabling high-resolution simulations of airflow, heat transfer, pollutant dispersion, and environmental interactions, CFD allows designers, engineers, and researchers to evaluate complex physical phenomena before construction or intervention [4, 10, 11]. This capability supports informed decision-making at both the building and urban scale, facilitating more accurate and climate-responsive planning and design processes [11].

Among the wide range of CFD applications in the built environment, fifteen core application areas have been identified as the most critical and impactful for these disciplines. To provide a clear and systematic overview, the relationship between these fifteen application areas and the capabilities of twenty-four CFD software platforms reviewed in this chapter has been summarized in two comprehensive tables. Table 5.1 presents a detailed matrix mapping each of the fifteen architectural applications to the twenty-four software tools, indicating which platforms support each application. Similarly, Table 5.2 provides a corresponding matrix for urban-scale applications. Together, these tables offer a structured reference point for practitioners and researchers, highlighting the strengths, limitations, and specialization areas of each software package in relation to the key challenges of architecture and urban planning.

These comparative summaries serve as the foundation for readers to identify the most appropriate CFD tools for specific design objectives, project scales, and interdisciplinary workflows.

5.3.1 Building Scale—Fifteen Key CFD Applications

#1. Natural Ventilation—Optimization of airflow pathways for passive cooling and indoor air quality improvement.

#2. Indoor Thermal Comfort—Evaluating and enhancing occupant comfort through precise thermal simulations.

#3. Windcatcher Design—Assessing the performance of traditional and modern windcatchers for passive ventilation.

#4. Window and Opening Optimization—Determining optimal placement and sizing of windows and openings to improve ventilation efficiency.

#5. Indoor Temperature Distribution—Simulating temperature gradients within enclosed spaces.

#6. HVAC System Design—Optimizing mechanical heating, ventilation, and air conditioning systems.

#7. Smoke Control in Fire Safety—Modeling smoke movement to support fire safety planning and emergency evacuation strategies.

#8. Fire Suppression System Design—Evaluating fire suppression methods using fluid dynamic simulations.

Table 5.1 Comparative analysis of 24 CFD software platforms for 15 key architectural applications

Software/Applications	#1	#2	#3	#4	#5	#6	#7	#8	#9	#10	#11	#12	#13	#14	#15
1. ANSYS Fluent	✓	✓	✓	✓	✓	✓	✓	✓	✗	✓	✓	✗	✓	✗	✓
2. ANSYS CFX	✓	✓	✓	✓	✓	✓	✓	✓	✗	✓	✓	✗	✓	✗	✓
3. GAMBIT	✗	✗	✗	✗	✗	✗	✗	✗	✗	✗	✗	✗	✗	✗	✗
4. ICEM-CFD	✗	✗	✗	✗	✗	✗	✗	✗	✗	✗	✗	✗	✗	✗	✗
5. TGrid / Fluent Meshing	✗	✗	✗	✗	✗	✗	✗	✗	✗	✗	✗	✗	✗	✗	✗
6. CFD-POST	✗	✗	✗	✗	✗	✗	✗	✗	✗	✗	✗	✗	✗	✗	✗
7. STAR-CCM +	✓	✓	✓	✓	✓	✓	✓	✓	✓	✓	✓	✗	✓	✗	✓
8. OpenFOAM	✓	✓	✓	✓	✓	✓	✓	✓	✗	✓	✓	✓	✓	✗	✓
9. Flow-3D	✗	✗	✗	✗	✗	✗	✗	✗	✗	✗	✗	✗	✗	✗	✗
10. COMSOL Multiphysics	✓	✓	✓	✓	✓	✓	✗	✗	✓	✓	✓	✓	✓	✓	✓
11. NUMECA	✗	✗	✗	✗	✗	✗	✗	✗	✗	✗	✗	✗	✗	✗	✗
12. AVL FIRE	✗	✗	✗	✗	✗	✗	✗	✗	✗	✗	✗	✗	✗	✗	✗
13. CFD++	✓	✓	✗	✗	✓	✓	✗	✗	✗	✓	✓	✗	✗	✗	✓
14. PumpLinx	✗	✗	✗	✗	✗	✗	✗	✗	✗	✗	✗	✗	✗	✗	✗
15. Autodesk CFD	✓	✓	✗	✓	✓	✓	✓	✓	✗	✓	✓	✗	✓	✓	✓
16. DesignBuilder CFD	✓	✓	✓	✓	✓	✓	✗	✗	✗	✓	✓	✓	✓	✓	✓
17. ENVI-met	✓	✓	✓	✗	✓	✗	✗	✗	✓	✓	✗	✓	✓	✓	✓
18. SimScale	✓	✓	✓	✓	✓	✓	✓	✓	✓	✓	✓	✓	✓	✓	✓
19. XFlow	✗	✗	✗	✗	✗	✗	✗	✗	✗	✗	✗	✗	✗	✗	✗
20. SU2	✓	✓	✓	✓	✓	✓	✗	✗	✗	✓	✓	✓	✓	✗	✓
21. HiFUN	✗	✗	✗	✗	✗	✗	✗	✗	✗	✗	✗	✗	✗	✗	✗
22. Palabos	✓	✓	✓	✓	✓	✓	✗	✗	✗	✓	✓	✓	✓	✗	✓
23. PowerFLOW	✗	✗	✗	✗	✗	✗	✗	✗	✓	✗	✗	✗	✗	✗	✗
24. Phoenix Integration	✓	✓	✓	✓	✓	✓	✗	✗	✗	✓	✓	✓	✓	✓	✓

#9. Acoustic Analysis—Studying sound propagation and noise control in building interiors.

#10. Wind Infiltration Analysis—Assessing unwanted air leakage and drafts within building interiors.

#11. Building Envelope Heat Transfer—Modeling heat exchange through the building envelope for energy efficiency analysis.

#12. Green Roof Design—Evaluating the impact of vegetated roofs on building energy performance and microclimate.

Table 5.2 Comparative analysis of 24 CFD software platforms for 15 key urban applications

Software/Applications	#1	#2	#3	#4	#5	#6	#7	#8	#9	#10	#11	#12	#13	#14	#15
1. ANSYS Fluent	✓	✓	✓	✓	✓	✓	✓	✓	✓	✓	✓	✓	✓	✓	✓
2. ANSYS CFX	✓	✓	✓	✓	✓	✓	✓	✓	✓	✓	✓	✓	✓	✓	✓
3. GAMBIT	✗	✗	✗	✗	✗	✗	✗	✗	✗	✗	✗	✗	✗	✗	✗
4. ICEM-CFD	✗	✗	✗	✗	✗	✗	✗	✗	✗	✗	✗	✗	✗	✗	✗
5. TGrid / Fluent Meshing	✗	✗	✗	✗	✗	✗	✗	✗	✗	✗	✗	✗	✗	✗	✗
6. CFD-POST	✗	✗	✗	✗	✗	✗	✗	✗	✗	✗	✗	✗	✗	✗	✗
7. STAR-CCM +	✓	✓	✓	✓	✓	✓	✓	✓	✓	✓	✓	✓	✓	✓	✓
8. OpenFOAM	✓	✓	✓	✓	✓	✓	✓	✓	✓	✓	✓	✓	✓	✓	✓
9. Flow-3D	✗	✗	✗	✗	✗	✗	✗	✓	✗	✗	✗	✗	✗	✗	✗
10. COMSOL Multiphysics	✓	✓	✓	✓	✓	✓	✓	✓	✓	✓	✓	✓	✓	✓	✓
11. NUMECA	✗	✗	✗	✗	✗	✗	✗	✗	✗	✗	✗	✗	✓	✗	✗
12. AVL FIRE	✗	✗	✗	✗	✗	✗	✗	✗	✗	✗	✗	✗	✗	✗	✗
13. CFD++	✓	✓	✓	✓	✓	✓	✓	✗	✓	✓	✓	✓	✓	✓	✓
14. PumpLinx	✗	✗	✗	✗	✗	✗	✗	✗	✗	✗	✗	✗	✗	✗	✗
15. Autodesk CFD	✓	✓	✗	✗	✗	✓	✓	✗	✓	✓	✓	✓	✓	✓	✓
16. DesignBuilder CFD	✓	✓	✓	✗	✗	✓	✓	✗	✓	✓	✓	✗	✓	✓	✓
17. ENVI-met	✓	✓	✓	✓	✓	✓	✓	✗	✓	✓	✗	✗	✗	✓	✓
18. SimScale	✓	✓	✓	✓	✓	✓	✓	✓	✓	✓	✓	✓	✓	✓	✓
19. XFlow	✓	✓	✗	✗	✗	✗	✓	✗	✓	✗	✗	✗	✓	✗	✗
20. SU2	✓	✓	✓	✓	✓	✓	✓	✗	✓	✓	✓	✗	✓	✓	✓
21. HiFUN	✗	✗	✗	✗	✗	✗	✗	✗	✗	✗	✗	✗	✗	✗	✗
22. Palabos	✓	✓	✓	✓	✓	✓	✓	✗	✓	✓	✓	✗	✓	✓	✓
23. PowerFLOW	✓	✓	✗	✓	✓	✗	✓	✗	✓	✓	✗	✗	✓	✓	✗
24. Phoenix Integration	✓	✓	✓	✓	✓	✓	✓	✓	✓	✓	✓	✓	✓	✓	✓

#13. Material Impact on Energy Use—Studying the thermal behavior of different construction materials.

#14. Solar Radiation Optimization—Analyzing solar gain and shading strategies to enhance energy performance.

#15. Semi-Open Space Airflow—Investigating airflow dynamics in semi-open spaces such as courtyards and patios.

The results presented in Table 5.1 highlight notable differences in the coverage of architectural applications across various CFD software platforms. Tools such as DesignBuilder CFD, ENVI-met, SimScale, and COMSOL Multiphysics are highly aligned with architectural design needs. These platforms are optimized for tasks such as natural ventilation, indoor thermal comfort, solar radiation analysis, and the evaluation of sustainable design features like green roofs and semi-open spaces. Their seamless integration with BIM workflows makes them particularly suitable for architects and designers during the early and intermediate design stages.

In contrast, general-purpose CFD platforms such as ANSYS Fluent, ANSYS CFX, STAR-CCM+, and OpenFOAM provide advanced capabilities for accurate fluid flow and heat transfer simulations. However, due to their complexity and the need for specialized expertise, they are more appropriate for research projects or complex simulations like fire dynamics, smoke control, and HVAC system design in large buildings. Supporting tools such as GAMBIT, ICEM-CFD, and TGrid are not used for direct simulation but serve specifically for mesh generation and pre-processing.

Finally, industrial and domain-specific software such as AVL FIRE, NUMECA, and Flow-3D have limited relevance to architectural design, being primarily developed for sectors like automotive, hydraulics, or turbomachinery. These tools are only applied in specialized cases, such as modeling water flow in unique structures or urban acoustics. Consequently, the choice of CFD software should be based on project scope and design objectives, with architecture-focused tools used for building performance and sustainable design, while general-purpose platforms are reserved for high-fidelity research and safety engineering tasks.

5.3.2 Urban Scale—Fifteen Key CFD Applications

#1. Urban Wind Flow Analysis—Understanding airflow patterns within complex urban geometries.

#2. Pedestrian Wind Comfort—Evaluating wind effects on pedestrian comfort and safety.

#3. Urban Heat Island Simulation—Modeling heat accumulation and dissipation in dense urban areas.

#4. Air Pollution Modeling—Tracking pollutant dispersion and concentration levels in urban environments.

#5. Traffic Emission Dispersion—Simulating the spread of vehicular emissions and their impact on urban air quality.

#6. Neighborhood Natural Ventilation—Designing natural ventilation strategies at the district or neighborhood scale.

#7. Building Shadow Analysis—Studying the shading effects of tall buildings on surrounding areas.

#8. Urban Flood Management—Modeling stormwater runoff and flood risk mitigation strategies.

#9. Street Canyon Wind Flow—Analyzing airflow within narrow streets and high-density urban corridors.

#10. Public Space Optimization—Designing parks and public areas for enhanced thermal comfort and airflow.

#11. Transit Hub Ventilation—Simulating ventilation in subway stations, terminals, and other public transport hubs.

#12. Smoke and Toxic Gas Dispersion—Modeling hazardous gas dispersion for emergency planning and safety design.

#13. Urban Wind Load Analysis—Assessing wind-induced forces on urban structures for stability and safety.

#14. Pedestrian Thermal Comfort—Evaluating thermal comfort for pedestrians in outdoor environments.

#15. Sustainable Urban Planning Evaluation—Assessing urban design strategies for energy efficiency and environmental sustainability.

The analysis of Table 5.2 reveals that CFD software platforms for urban-scale simulations can be categorized into three main groups. The first group consists of comprehensive, multiphysics platforms such as ANSYS Fluent, STAR-CCM + , OpenFOAM, and SimScale, which support nearly all fifteen key urban applications. These include urban wind flow modeling, air pollution analysis, urban flood management, neighborhood-scale natural ventilation design, and pedestrian thermal comfort assessment. Owing to their advanced capabilities, these tools are particularly well-suited for large-scale urban projects and master planning initiatives that demand high-fidelity, multi-physics modeling, including sophisticated turbulence, heat transfer, and pollutant dispersion simulations. Supporting tools such as GAMBIT, ICEM-CFD, and TGrid are exclusively used for mesh generation and pre-processing and do not perform CFD simulations themselves.

The second group includes specialized urban-focused tools such as ENVI-met, Palabos, and PowerFLOW, which place a strong emphasis on urban microclimate, urban heat island analysis, pedestrian comfort, and ventilation of outdoor spaces. For instance, ENVI-met is highly effective for sustainable urban planning and public space design due to its advanced models for simulating vegetation effects, solar radiation, and the thermal influence of urban materials on the microclimate. Similarly, lattice Boltzmann method (LBM)-based platforms like Palabos are well-suited for airflow simulations in dense and complex urban geometries, particularly in studies involving air pollution dispersion and natural ventilation strategies at the neighborhood scale. However, compared to the first group, these tools are less versatile and have more limited coverage in industrial or infrastructure-related applications.

The third group comprises niche and domain-specific software designed for highly specialized urban challenges. Flow-3D, for example, is particularly effective for urban flood modeling and stormwater drainage design. Other industrial platforms, such as AVL FIRE and NUMECA, are generally unsuitable for urban planning and are used only in specific cases, such as wind load simulations on tall structures. Supporting tools like GAMBIT, ICEM-CFD, and TGrid serve exclusively for mesh generation and pre-processing and do not perform CFD simulations themselves. Consequently, the selection of an appropriate CFD tool should be guided by project objectives and study scale: comprehensive multiphysics platforms are ideal for integrated urban modeling, specialized software for microclimate and sustainable design studies, and niche tools for infrastructure-specific analyses.

5.4 Summary of Findings from This Comparative Analysis of CFD Software for Architectural and Urban Environmental Simulation

This chapter presents a comprehensive comparative review of 24 computational fluid dynamics (CFD) software packages, ranging from widely used commercial solvers to advanced open-source and cloud-based platforms. The analysis includes established tools such as ANSYS Fluent, CFX, and COMSOL Multiphysics, as well as emerging platforms like SimScale, SU2, and Palabos. Each software is evaluated in terms of its meshing capabilities, turbulence and multiphase modeling, coupling with heat transfer and chemical reactions, user interface structure, and support for transient or steady-state simulations.

Special attention is given to the applicability of these tools in the architecture, urban planning, and building engineering sectors. This includes CFD applications in HVAC design, outdoor wind comfort, daylight performance, and urban microclimate analysis. Tools like ENVI-met, DesignBuilder CFD, Autodesk CFD, and COMSOL are specifically examined for their relevance to integrated building simulation. In addition, the chapter highlights recent technological trends such as GPU acceleration, adaptive mesh refinement (AMR), lattice Boltzmann methods, and hybrid structured-unstructured meshing techniques. This chapter aims to support researchers, engineers, and designers in selecting the most appropriate CFD platform for their specific interdisciplinary modeling needs.

5.5 Conclusion: Toward Informed CFD Tool Selection for the Built Environment

The careful selection of computational fluid dynamics (CFD) software has become an essential part of successful architectural and urban environmental simulation as the built environment gets more complex and climate-responsive design becomes a global necessity. 24 CFD tools with differing degrees of expertise, computational performance, and user accessibility have been thoroughly compared in this chapter. By exploring the specific functionalities of each platform—ranging from mesh generation and turbulence modeling to multi-physics coupling and post-processing capabilities—we have aimed to offer a practical guide to professionals and researchers working at the intersection of simulation and sustainable design.

This comparative analysis makes it clear that no CFD program can be said to be universally better than another. The specific requirements of the project, whether they include large-scale urban wind simulations, high-fidelity modeling of indoor ventilation and contaminant dispersion, or integrated evaluations of energy use, daylighting, and comfort within building envelopes, ultimately dictate the most efficient platform. Open-source tools like OpenFOAM and SU2 continue to improve in flexibility, affordability, and community-driven innovation, while commercial platforms like ANSYS Fluent and COMSOL Multiphysics provide strong multi-physics and customization features. SimScale and other cloud-based systems are becoming more popular due to their scalability and accessibility, particularly among interdisciplinary teams and early-career researchers.

Crucially, CFD techniques are useful in architectural and urban contexts for strategic as well as technical reasons. Simulation is used as a tool for decision-making that guides design interventions, retrofitting plans, and policy decisions as cities struggle with excessive heat, air pollution, and energy waste. A new paradigm of data-driven design where airflow, heat transfer, and climate responsiveness are ingrained from the very beginning of the architectural process is indicated by the integration of CFD with energy analysis software, BIM workflows, and environmental performance tools.

This chapter concludes by highlighting the need to comprehend the subtleties of CFD platforms, including their computational capacity, usability, interoperability, and applicability to geographical and environmental issues. Our ability to evaluate, integrate, and apply simulation technologies in ways that promote robust, effective, and human-centered design must develop along with the technology.

References

1. Kochkov, D., Smith, J. A., Alieva, A., Wang, Q., Brenner, M. P., & Hoyer, S. (2021). Machine learning–accelerated computational fluid dynamics. *Proceedings of the National Academy of Sciences, 118*(21), Article e2101784118.
2. Chen, X., Wang, Z., Deng, L., Yan, J., Gong, C., Yang, B., & Pang, Y. (2024). Towards a new paradigm in intelligence-driven computational fluid dynamics simulations. *Engineering Applications of Computational Fluid Mechanics, 18*(1), 2407005.
3. Idrissi, A., El Mghari, H., & El Amraoui, R. (2025). CFD simulation advances in urban aerodynamics: Accuracy, validation, and high-rise building applications. *Results in Engineering*, 105148.
4. Chu, R., & Wang, K. (2025). CFD in urban wind resource assessments: A review. *Energies, 18*(10), 2626.

5. Caron, C., Lauret, P., & Bastide, A. (2025). Machine learning to speed up computational fluid dynamics engineering simulations for built environments: A review. *Building and Environment, 267*, Article 112229.
6. Ahmad, F., Majumder, D., Ranjit, R., Gupta, A., & Manhart, M. (2025). Preliminary study on the spread of air-borne pollutants in urban environment: A CFD simulation approach. *Scientific Reports, 15*(1), 18836.
7. Mao, R., Lan, Y., Liang, L., Yu, T., Mu, M., Leng, W., & Long, Z. (2025). Rapid CFD prediction based on machine learning surrogate model in built environment: A review. *Fluids, 10*(8), 193.
8. Wijesooriya, K., Mohotti, D., Lee, C.-K., & Mendis, P. (2023). A technical review of computational fluid dynamics (CFD) applications on wind design of tall buildings and structures: Past, present and future. *Journal of Building Engineering, 74*, Article 106828.
9. Lee, D., Barquilla, C. A. M., & Lee, J. (2025). Analyzing dispersion characteristics of fine particulate matter in high-density urban areas: A study using CFD simulation and machine learning. *Land, 14*(3), 632.
10. Blocken, B. (2015). Computational fluid dynamics for urban physics: Importance, scales, possibilities, limitations and ten tips and tricks towards accurate and reliable simulations. *Building and Environment, 91*, 219–245.
11. Altalbawy, F. M. A., Alkhayyat, A., Jhala, R., Yadav, A., Ramachandran, T., Shankhyan, A., & Sinitsin, V. V. (2025). Computational fluid dynamic and machine learning modeling of nanofluid flow for determination of temperature distribution and models comparison. *Scientific Reports, 15*(1), 19073.

Chapter 6
From Simulation to Design Intelligence: Lessons, Reflections, and Pathways Ahead for Applied CFD in Architecture and Urbanism

Abstract This concluding chapter of *Applied CFD: A Practical Guide for Architects and Planners* summarizes the theoretical and practical journey of the book by showing how computational fluid dynamics (CFD) helps to bridge the gap between scientific accuracy and creative design. It examines how physics, numerical modelling, experimental verification, and software application are combined to show how CFD enables planners and architects to create environmentally intelligent designs. Lessons learned emphasise the importance of calibration, validation, and balanced fidelity in modelling, while underscoring CFD's role as both a design exploration and communication tool. The chapter looks at the future of applied CFD by imagining how it will merge with digital twins, AI-assisted processes, and predictive modelling for climate-responsive design. It emphasizes the increasing convergence of data-driven design and environmental stewardship and promotes the democratization of simulation through open-source and cloud-based tools. In the end, CFD is presented as a transformative language rather than just a technical technique, one that allows science and design to work together to create cities that are healthier, more adaptable, and sustainable.

Keywords Computational Fluid Dynamics · CFD · Architecture · Urban design · Environmental simulation · Digital twins · AI-Assisted Design · Climate-Responsive Cities

6.1 Integrating Knowledge: From Physics to Design Practice

From a technical field based on fluid mechanics, computational fluid dynamics (CFD) has developed into an artistic tool of architectural investigation. From the fundamental physics of flow and energy conservation (Chaps. 1 and 2) to their numerical discretization and computation (Chap. 3), experimental validation in controlled settings (Chap. 4), and, lastly, their practical implementation in software platforms

J. Shaeri and A. Cheshmehzangi, *Applied CFD*, Urban Sustainability,
https://doi.org/10.1007/978-981-95-8505-2_6

utilized by planners and designers (Chap. 5), the preceding chapters outline this intellectual progression. The potential of applied CFD to connect the concrete reality of the built environment with the abstract accuracy of mathematics is encapsulated in this process.

The conservation laws of mass, momentum, and energy—expressed in Chap. 1—are the theoretical foundation of CFD. These concepts give us a physics-based understanding of comfort, efficiency, and safety by describing how heat and air flow through and around structures. These are not just equations to planners and architects; they are representations of how space reacts to climate and breathes. This framework was expanded to the irregular geometries of architecture, such as roofs, facades, and urban canyons, where air and heat act quite differently from idealized laboratory settings. This was accomplished in Chap. 2 by adapting these governing equations to generalized coordinate systems. As a result, universal physics is translated into a form that is sensitive to local shape and materiality.

Chapter 3 then provided the computational bridge. The continuous flow of air was discretized into calculable cells and moments via the finite volume method (FVM) and its auxiliary algorithms, which converted theoretical formulations into solvable models. The intangible became quantifiable within this numerical grid. Coupled heat-transfer equations, near-wall treatments, and turbulence modelling provided the rigor needed to describe environmental behavior at architectural resolution. Yet this chapter also underscored the iterative nature of modelling: no computation is final without calibration; no result is meaningful without validation.

The crucial conversation between the digital and the physical was demonstrated by that validation, which was brought to life through wind tunnel testing in Chap. 4. By converting abstract fields of pressure and velocity into empirically validated patterns, experimental analysis made it possible to assess the validity of CFD. The book reiterated a fundamental fact by contrasting simulations with controlled airflow experiments: CFD enhances experimentation rather than replaces it. The foundation of modern performance-based design is the hybrid paradigm, which consists of computation leading experimentation and experiment correcting computation. It also represents a more general epistemology of architectural science, which holds that knowledge is developed by repeatedly comparing representation and reality.

Chapter 5, which grounded the theory and validation into real-world application, was the culmination of this journey. In this case, CFD platforms were contrasted as design tools rather than abstract tools, each with unique affordances for urban and architectural settings. Commercial software like ANSYS Fluent and COMSOL Multiphysics offer dependability, multi-physics integration, and interoperability with BIM systems; open-source platforms like OpenFOAM and SU2 promote transparency and adaptability, fostering parametric workflows and academic innovation; and cloud-based platforms like SimScale democratize simulation through accessibility and real-time collaboration. These technologies work together to create a dispersed ecology where creative design and accurate modelling coexist.

CFD is changed from an engineering specialization to a design philosophy by this interdisciplinary synthesis. Architects and planners are now interpreters of physical events, able to perceive wind patterns as spatial tales, rather than merely passive consumers of simulation tools. By showing the invisible processes of turbulence, convection, and diffusion that significantly influence ecological performance and human comfort, CFD helps to promote this literacy. Before laying a single brick, designers can test, fail, and improve concepts using iterative modeling–validation loops. The end effect is an emergent design culture in which environmental behavior is viewed as a design parameter rather than a limitation and intuition is supported by data.

At its best, applied CFD operates/functions as a compass and a microscope. It acts as a microscope, displaying minute details about pollution dispersion, heat gradients, and flow interactions. It acts as a compass, directing choices toward solutions that balance functionality and style. This dual purpose supports the main contention of the book, which holds that social and spatial responsibilities are inextricably linked to computational understanding of fluid behavior [1]. The process of modelling airflow across a courtyard involves more than just math; it involves designing for ecological balance, energy efficiency, and human experience.

The integration of physics and practice, therefore, is more than methodological—it is cultural. It represents the increasing understanding that urban planning and design are sciences of interaction rather than of isolation. When used carefully, CFD prepares the upcoming generation of designers to handle these relationships both creatively and statistically. It cultivates an attitude sensitive to feedbacks, systems, and change dynamics [2–5]. In this way, the intellectual path outlined in the first five chapters leads to epistemic maturity rather than technical mastery: the insight that comprehending the constructed environment necessitates both creative interpretation and analytical precision.

Building on this synthesis, the next chapter explores the future of applied CFD, focusing on how it will continue to develop as a scientific field and a design catalyst, influencing the development of performance-driven and climate-responsive architecture.

6.2 Lessons Learned: The Role of CFD in Architectural and Urbanism

The exploration of the preceding chapters emphasizes that CFD is both a science of accuracy and an art of interpretation. By describing, evaluating, and applying the approach to architectural and urban design contexts, each chapter adds a layer to this understanding. Several fundamental principles that improve CFD practice and instruction and make sure it complements rather than overshadows design purpose are revealed by these implementations.

The first and perhaps most critical lesson concerns model calibration and validation. CFD is still a numerical approximation of reality in spite of its mathematical precision. The wind tunnel setup experiments in Chap. 4 demonstrate the critical importance of empirical validation. The model moves from assumption to trust when digital simulations are compared to physical data, such as pressure coefficients, turbulence intensities, or velocity fields. Therefore, validation must be viewed by architects and planners as an ethical requirement rather than as an academic exercise. Whether it is measuring wind safety around tall buildings, assessing pedestrian comfort in urban canyons, or enhancing natural ventilation systems, a validated CFD model gives trust in its projections. On the other hand, if this phase is skipped, simulation could become conjecture. Hence, the discipline's credibility depends on maintaining this dialogue between the physical and the digital.

Finding a balance between computational pragmatism and modeling fidelity is the second important lesson. As discussed in Chap. 3, using sophisticated turbulence models like large eddy simulation (LES) or increasing mesh resolution improves accuracy but also increases processing requirements dramatically. The temptation to over-model is something that planners and architects must avoid. Simple steady-state Reynolds-averaged Navier–Stokes (RANS) models are frequently enough for early design considerations, such as orientation, massing, or fenestration. Only when design geometries have been finalized and comprehensive environmental certification has been sought may high-fidelity simulations be used. Understanding these trade-offs guarantees that CFD stays a tool for decision-making rather than a scholarly luxury. With the increasing availability of cloud-based solvers like SimScale and multi-physics tools like COMSOL, efficient use of computational resources necessitates intentional simplification without sacrificing design quality.

The third lesson involves understanding and respecting the assumptions within CFD. Laminar-turbulent transition models, boundary conditions, wall functions, and steady-state approximations are examples of simplifications that are present in every simulation. Even if the output seems visually impressive, it might be misleading if these assumptions are misunderstood or ignored. The comparative software assessment in Chap. 5 serves as a reminder to readers that CFD tools are intricate systems of interconnected equations and numerical solvers rather than mysterious black boxes. One of the major errors in the field is the over-reliance on visual outputs, like colorful velocity contours, without questioning solver parameters, grid independence, or convergence criteria. Intellectual humility is the lesson here: every visualization needs to be examined, every result questioned. While engineers should learn to explain the underlying uncertainties in a language that designers can understand, architects who use CFD should develop a sufficient level of numerical literacy to critically assess results.

From a pedagogical standpoint, CFD's greatest strength may be its educational value [6]. CFD becomes a tool for learning environmental behavior through simulation as opposed to memorization when it is incorporated into architectural courses. By experimenting with form, students may see how little design changes affect temperature, airflow, and pollution dispersion. Designers get to "see" invisible influences as a result of this iterative process, which develops environmental intuition. A theoretical basis for this literacy is established by Chap. 1's focus on the physics of airflow and Chap. 2's thorough description of coordinate systems and governing equations. Early design education that incorporates CFD fosters scientific thinking and creative experimentation, two abilities that are becoming more and more important in the age of climate-responsive design.

Finally, CFD serves as a communication bridge between disciplines. Architects, engineers, planners, and legislators may all visualize, discuss, and argue complex environmental issues. From micro-scale room ventilation to macro-scale metropolitan airflows, the simulations included in the book illustrate the narrative power of CFD. A well-built simulation turns invisible dynamics into shared design evidence, telling tales of motion, heat, and wind in addition to numbers. Accessible CFD platforms have extended this communicative potential outside of research labs and into design studios and planning offices, as demonstrated by the comparative assessments in the book. Design teams may match environmental performance with social, aesthetic, and regulatory goals when CFD is employed appropriately.

In conclusion, the lessons learned from the case studies and comparative analyses support the idea that the value of CFD is in developing sound judgment rather than making flawless forecasts. A balanced combination of scientific discipline, computational constraint, and creative intuition is required for the proper use of CFD. Furthermore, CFD will continue to be an essential tool as planners and architects tackle ever-more-complex urban problems; it directs design thinking through the movement of data, energy, and air, converting environmental consciousness into concrete spatial intelligence.

6.3 Pathways Ahead: Evolving Tools, Methods, and Interdisciplinary Workflows

An integrated ecosystem of digital information, environmental foresight, and design adaptability is quickly replacing the isolated technical application of applied CFD in architecture and planning. In the future, simulation will be a creative force in creating resilient, sustainable, and intelligent cities rather than an afterthought or validation tool, thanks to the foundations this book lays by tying physics, computation, and practice together. To enable design systems to learn, forecast, and react in ways that reflect the dynamism of the built environment itself, CFD will need to be combined with AI, real-time sensing, and multi-physics modelling.

The integration of CFD with digital twins, parametric workflows, and AI-assisted modelling is a notable development currently in progress [7, 8]. The logical progression of CFD's reasoning from simulation to continuous feedback is represented by digital twins, which are real-time, data-driven virtual representations of buildings and urban systems. Designers and planners can develop adaptive systems that modify airflow and thermal predictions in response to shifting conditions by combining CFD with real-time environmental data [9]. For example, cities can use predictive analytics to reduce energy usage, modify ventilation schedules, or predict heat islands by combining CFD-based digital twins with building management systems. This approach is further improved by AI: machine learning algorithms can produce quick, approximative flow estimates for early-stage design by learning from hundreds of CFD simulations [10]. By significantly cutting down on computational time, these surrogate models allow architects to evaluate hundreds of design variations in the time that would have been needed for a single high-fidelity simulation. This convergence, in which generative design cycles integrate physics, performance, and form-finding into a single, coherent procedure, is already hinted at by parametric platforms like Grasshopper, Dynamo, and Ladybug Tools.

CFD's function broadens as it combines with intelligent systems, moving beyond performance analysis to include predictive resilience and thermal health. After a design has been created, CFD has historically been used to assess temperature and airflow distributions. CFD is repositioned as a proactive diagnostic tool in the following step. Planning approaches for climate-sensitive cities can benefit from predictive modelling, which can detect heat-vulnerable areas or ventilation bottlenecks before they become apparent. Global pandemic experiences have supported the idea that CFD may be used to model pathogen dispersion in interior environments and optimize ventilation methods for public safety in health-conscious design. By combining CFD with energy and weather models, it is possible to estimate microclimatic behavior across districts at the urban scale, which helps with decisions about building orientation, street design, and green infrastructure. CFD is elevated from an analytical technique to a strategic framework for environmental regulation by such predictive applications. In order to handle climate adaption, heat stress, and air quality in real time, the future city might rely on an ongoing CFD–data feedback loop that uses digital twins at the citywide scale.

Another important avenue is democratization, which is breaking down the exclusivity that formerly surrounded high-end simulation through open-source and cloud-based CFD. Research has already been transformed by open-source systems like OpenFOAM and SU2, which let users access, alter, and increase solver capabilities. Thanks to their continuous development, intuitive interfaces, and streamlined setup procedures, architects and students may now run simulations without needing extensive programming experience. Further, browser-based simulation environments that run on distant servers are provided by cloud-based systems like SimScale or Ansys Discovery Live, which remove the hardware barrier for smaller studios. This democratization reflects a broader cultural change: simulation is now a component of a collective design intelligence that interdisciplinary teams can access rather than being the sole purview of professional engineers. The next generation of designers will be able to approach environmental performance as a creative process rather than a technical burden if CFD is integrated into design education and professional practice.

However, new obligations are also brought about by this accessibility. The requirement for interpretive literacy increases as tools become more user-friendly. Democratizing CFD should not equate to simplifying the science. Training should prioritize comprehension over automation, making sure that users continue to question solver integrity, boundary conditions, and model assumptions. New educational frameworks that teach CFD as an integrated language of environmental design—one that blends scientific rigor, computational proficiency, and architectural judgment—must be developed by academic institutions and professional associations. In this way, the development of CFD is similar to that of architecture in that it has kept its ethical and intellectual depth while broadening its scope of knowledge.

The final and perhaps most transformative frontier is the convergence of data-driven design, machine learning, and computational environmental science. Machine learning adds probabilistic reasoning to traditional CFD, which is based on deterministic physics. These paradigms work together to create hybrid models that are based on physical law but can also learn from data. The use of such hybrid "physics-informed neural networks" to speed up CFD solvers, lessen mesh reliance, and bridge gaps in sparse environmental information is already being investigated. This could lead to systems in urban design that can simulate wind behavior in changing neighborhoods or forecast the thermal comfort of entire precincts without the need for complete meshing or boundary configuration. The potential is enormous: designers can create self-improving urban models by incorporating learning algorithms into simulation workflows, which will allow them to constantly improve models as new environmental data becomes available.

Collaboration amongst architects, computer scientists, environmental engineers, and urban ecologists is also encouraged by this interdisciplinary convergence. As data pipelines connect sensors, simulations, and spatial models into cohesive design ecosystems, the formerly distinct borders of expertise start to become less distinct. This convergence is best illustrated by the emergence of "computational environmental science," a new area that blends spatial modelling, AI, and environmental analytics to produce useful information for sustainable design. In this case, CFD is a part of a larger constellation of prediction technologies that address climate adaptation, resilience, and equity issues.

In the years to come, the most effective applications of CFD will not be the most intricate but rather the most integrated, where simulation becomes an embedded intelligence in the design process and numerical accuracy influences human-centered design [11]. In addition to improved algorithms, the future of applied CFD also depends on improved interdisciplinary cooperation, human–machine interaction, and the balance between environmental responsibility and creative design. To create an environment that breathes, adapts, and survives with more intelligence and care, architects and planners must use these developing capabilities to create more than just a simulation of air and heat.

6.4 Concluding Reflections: Architects and Planners in the Age of Simulation

This book's evolution shows how our understanding and design of the built world have undergone a significant shift. A more comprehensive examination of the interplay between science and creativity, accuracy and imagination, has emerged from what started as a technical investigation of airflows, heat transport, and turbulence. CFD, or applied CFD, has shown itself to be a useful tool for analysis as well as a conduit between fields, languages, and perspectives. It has shown that architecture involves more than just forming space; it also involves coordinating the unseen forces that keep life alive there.

Today, planners and architects are at the cusp of a new paradigm in design, one that is characterized by the dynamic behavior of surroundings rather than the static geometry of form. Air, heat, and energy flows can be simulated and visualized, giving designers a sense of atmospheric authorship. The invisible becomes observable, quantifiable, and modifiable via CFD. It redefines beauty via performance and turns intuition into well-informed decision-making. The twenty-first century is characterized by the mastery of energy and flow, whereas the twentieth century celebrated the triumph of materials and structure. The management of the physical systems that provide a building or city with its environmental pulse is now part of the designer's skill, which goes beyond the obvious.

However, as this book has demonstrated, there are responsibilities associated with the power of simulation. The simplicity of digital modelling runs the risk of transforming data into ornamentation and complexity into spectacle. Being humble in front of the physics it interprets is a prerequisite for true environmental design. Every pressure contour or colorful velocity field is a hypothesis—an approximate representation of a living system whose complete complexity is never fully captured. Therefore, CFD should be viewed by planners and architects as a tool for inquiry, iteration, and learning rather than as a source of assurance. In this way, the field goes back to its scientific foundations: testing, experimentation, and knowledge evolution.

In its most profound form, CFD reconnects design with nature rather than separating it. By examining heat and airflow, we can once again see that cities and structures are permeable systems that exchange energy with their environment. Every simulation reaffirms that balance between natural and built systems leads to comfort, safety, and sustainability. CFD becomes a tool of stewardship when it is guided by ecological understanding, allowing designs that complement rather than defy climate, terrain, and culture. It serves as a reminder that technology by itself cannot accomplish sustainability; rather, our application of it must be intelligent.

The challenge for the upcoming generation of designers will be to interpret CFD as well as use it, converting numerical accuracy into spatial intelligence. Art and science come together in this act of translation. Living cities are not made of numbers alone; meaning is created when those numbers are converted into choices regarding ventilation, shade, proportion, and material. The planner turns into a curator of movement, and the architect into an interpreter of flow. Relationships between people and air, heat and light, and comfort and conservation are all things they must coordinate. In the era of simulation, this is the language of design: conscience-driven computing and data infused with empathy.

In terms of the future, applied CFD is a starting point rather than a conclusion—a gateway to the next phase of design thinking. Cities will soon become self-learning creatures that can dynamically adjust to weather, energy, and human demands thanks to the convergence of AI, digital twins, and real-time environmental monitoring. In this scenario, simulation will be a part of everyday operations, continuously improving how buildings function and how people perceive them, rather than serving as a preliminary to construction. Planners and architects will work with the climate to create feedback systems that change over time, rather than just reacting to it.

Therefore, the era of simulation is also the era of co-evolution, i.e., between science and design, between environmental intelligence and human creativity. CFD teaches us that comprehension, not control, is the key to advancement; conversing with the environment, not controlling it, is the key to progress. This book's ending offers continuation rather than closure—an invitation to continue investigating, trying new things, and dreaming. Because every whirlpool and flow pattern serves as a reminder that cities are living systems, always changing and evolving, much like the air that passes through them. Designing with that awareness means designing with vision, purpose, and consideration for the common environment that binds us all.

References

1. Zhai, Z. (John). (2020). *Computational fluid dynamics for built and natural environments.* Springer Singapore. https://doi.org/10.1007/978-981-32-9820-0
2. Cheshmehzangi, A. (2016). Multi-spatial environmental performance evaluation towards integrated urban design: a procedural approach with computational simulations. *Journal of Cleaner Production, 139*, 1085–1093. https://doi.org/10.1016/j.jclepro.2016.08.151
3. Cheshmehzangi, A., Zhu, Y., & Li, B. (2010). Integrated design approach-urban design for sustainability. In *5th International Conference on Responsive Manufacturing-Green Manufacturing (ICRM 2010)* (pp. 241–247). IET.
4. Shaeri, J., Mahdavinejad, M., & Pourghasemian, M. H. (2022). A new design to create natural ventilation in buildings: Wind chimney. *Journal of Building Engineering, 59*, 105041. https://doi.org/10.1016/j.jobe.2022.105041
5. Cho, S., Cho, S., Jung, S., & Kim, J. (2025). Locating urban area heat waves by combining thermal comfort index and computational fluid dynamics simulations: The optimal placement of climate change infrastructure in a Korean City. *Climate, 13*(6), 113.
6. Eldredge, J. D., Senocak, I., Dawson, P., Canino, J., Liou, W. W., LeBeau, R., Hitt, D. L., Rumpfkeil, M. P., & Cummings, R. M. (2014). A best practices guide to CFD education in the undergraduate curriculum. *International Journal of Aerodynamics, 4*(3–4), 200–236.
7. Diniz, P., Grimm, B., Garcia, F., Fayad, J., Ley, C., Mouton, C., Oeding, J. F., Hirschmann, M. T., Samuelsson, K., & Seil, R. (2025). Digital twin systems for musculoskeletal applications: A current concepts review. *Knee Surgery, Sports Traumatology, Arthroscopy, 33*(5), 1892–1910.
8. Adewoyin, M. A., Ogunnowo, E. O., Fiemotongha, J. E., Igunma, T. O., & Adeleke, A. K. (2024). Advances in AI-based optimization for multiphysics fluid dynamics in product design engineering.
9. Shaeri, J., Mahdavinejad, M., Vakilinejad, R., Bazazzadeh, H., & Monfared, M. (2023). Effects of sea-breeze natural ventilation on thermal comfort in low-rise buildings with diverse atrium roof shapes in BWh regions. *Case Studies in Thermal Engineering, 41*, 102638.
10. Mao, R., Lan, Y., Liang, L., Yu, T., Mu, M., Leng, W., & Long, Z. (2025). Rapid CFD prediction based on machine learning surrogate model in built environment: A review. *Fluids, 10*(8), 193.
11. Bereziartua-Gonzalez, L., Retegi, A., & Ukar, O. (2025). Human-centered integration of small wind turbines in urban environments: A semi-systematic review from an industrial design perspective. *Frontiers in Sustainable Cities, 7*, 1561894.

MIX
Papier aus verantwortungsvollen Quellen
Paper from responsible sources
FSC® C105338

If you have any concerns about our products,
you can contact us on
ProductSafety@springernature.com

In case Publisher is established outside the EU,
the EU authorized representative is:
Springer Nature Customer Service Center GmbH
Europaplatz 3, 69115 Heidelberg, Germany

Printed by Libri Plureos GmbH
in Hamburg, Germany